# Mathematicians and Education Reform
1989–1990

Conference Board of the Mathematical Sciences

# CBMS

## Issues in Mathematics Education

Volume 2

# Mathematicians and Education Reform 1989–1990

Naomi D. Fisher
Harvey B. Keynes
Philip D. Wagreich
Editors

**American Mathematical Society**
Providence, Rhode Island
in cooperation with
**Mathematical Association of America**
Washington, D. C.

This volume was compiled by the Mathematicians and Education Reform Network (MERNetwork). Its activities are supported by National Science Foundation Grant TPE-8850359.

---

**Library of Congress Cataloging-in-Publication Data**
Mathematicians and education reform 1989–1990/Naomi D. Fisher, Harvey B. Keynes, Philip D. Wagreich, editors.
    p. cm.—(Issues in mathematics education, ISSN 1047-398X; v. 2)
    At head of title: CBMS, Conference Board of the Mathematical Sciences.
    Includes bibliographical references.
    ISBN 0-8218-3502-5
    1. Mathematics—Study and teaching—Congresses.  I. Fisher, Naomi D.  II. Keynes, Harvey.  III. Wagreich, P. (Philip), 1941- .  IV. Conference Board of the Mathematical Sciences.  V. Series.
QA11.A1M2775    1991                                                91-15768
510'.71—dc20                                                            CIP

---

**Copying and reprinting.** Individual readers of this publication, and nonprofit libraries acting for them, are permitted to make fair use of the material, such as to copy an article for use in teaching or research. Permission is granted to quote brief passages from this publication in reviews, provided the customary acknowledgment of the source is given.

    Republication, systematic copying, or multiple reproduction of any material in this publication (including abstracts) is permitted only under license from the American Mathematical Society. Requests for such permission should be addressed to the Manager of Editorial Services, American Mathematical Society, P.O. Box 6248, Providence, Rhode Island 02940.

Copyright © 1991 by the American Mathematical Society. All rights reserved.
Printed in the United States of America.
The American Mathematical Society retains all rights
except those granted to the United States Government.
The paper used in this book is acid-free and falls within the guidelines
established to ensure permanence and durability. ∞
This publication was typeset using $\mathcal{A}_{\mathcal{M}}\mathcal{S}$-TEX,
the American Mathematical Society's TEX macro system.

10 9 8 7 6 5 4 3 2 1    96 95 94 93 92 91

# Contents

Foreword — vii

## Projects

Improving Mathematics Education through School-Based Change — 3
*Judith Mumme and Julian Weissglass*

Diagnostic Testing: One Link between University and High School Mathematics — 25
*Alfred Manaster*

Creativity: Nature or Nurture? A View in Retrospect — 39
*Arnold E. Ross*

Equity and Excellence in the University of Minnesota Talented Youth Mathematics Program (UMTYMP) — 85
*Harvey B. Keynes*

A Report on an Entry Level Math Program — 97
*Elias Toubassi*

## Issues and Reactions

Pros and Cons of Teaching Mathematics Via a Problem-Solving Approach — 115
*Bert Fristedt*

Obstacles to Change: The Implications of the National Council of Teachers of Mathematics (NCTM) *Standards* for Undergraduate Mathematics — 119
*T. Christine Stevens*

The Role of Teachers in Mathematics Education Reform — 127
*Joseph G. Rosenstein*

Teaching to Love Wisdom — 137
*Christopher Cotter and Igor Szczyrba*

Teacher Networking: A Corollary of Junior Mathematics Prognostic Testing 145
*Frank L. Gilfeather and Nancy A. Gonzales*

Using Technology for Teaching Mathematics 161
*Carl Swenson*

# Foreword

This second volume of papers on the work of mathematicians in education reform continues to underscore the range and intensity of mathematicians' involvement in educational reform efforts. As with volume I of *Mathematicians and Education Reform, Proceedings of the July* 6–8, 1988 *Workshop*, we have organized the articles into two sections, "Projects" and "Issues and Reactions." Making the final call for categorizing an article is sometimes a subtle choice. While projects are undertaken to address a problem or need, the thoughtful deliberation of an issue includes considering how to bring about change. The dynamics of analyzing a problem, putting ideas into action, and evaluating the outcome is a very vital and exciting part of the educational reform process.

We have included a variety of activities as a means to emphasize the range of mathematicians' interests in education. There are two reasons for this approach. First, it highlights that there are many ways for mathematicians to contribute to improving mathematics education. And second, it emphasizes that reform must be pursued on many fronts, not through a single avenue. Finally, amid this diversity of activities, it is possible to discern common underlying themes which reveal fundamental aspects of educational work.

Just as one makes a long-time professional commitment as a mathematician, serious educational work requires a sustained investment of time and energy. Characteristically, immersion in one aspect of education yields unexpected insights and opportunities for improving other parts of the educational picture. Deepening understanding is acquired through experience in formulating, testing, and refining—perhaps even abandoning—original notions. What may be most novel about the experience of educational work is the recognition that college and university mathematics faculty are part of a much larger community of professional educators and teachers of mathematics. Defining shared goals and finding ways to work together to accomplish these goals is at the core of successful reform in mathematics education.

Naomi D. Fisher
University of Illinois at Chicago

# Projects

# Improving Mathematics Education through School-Based Change

JUDITH MUMME AND JULIAN WEISSGLASS

## INTRODUCTION

Educators are being challenged to make fundamental changes in mathematics instruction and in schools themselves. (Carnegie Task Force on Teaching as a Profession (1986), Goodlad (1984), National Council of Teachers of Mathematics (1989), National Research Council (1989), National Science Board (1983).) The difficulties of achieving and sustaining educational change and curriculum reform, however, have been well documented. Fullan (1982), for example, states "There's no need to dwell on the fact that the vast majority of curriculum development and other educational change adoptions in the 1960's and 70's did not get implemented in practice, even where implementation was desired." The current reform effort started in 1983 with the publication of *Educating Americans for the 21st century* (National Science Board (1983)). Since then many dedicated teachers have made changes in their classrooms. Yet the overall approach to mathematics instruction remains much the same. It "continues to be dominated by teacher explanations, chalkboard presentations, and reliance on textbooks and workbooks. More innovative forms of instruction—such as those involving small group activities, laboratory work, and special projects—remain disappointingly rare" (Dossey, Mullis, Lindquist, and Chambers (1988)). It is our belief that fundamental change in mathematics instruction will require new professional development strategies—ones that address schools as an interconnected whole and provide emotional as well as intellectual support to teachers.

Project TIME (Teachers Improving Mathematics Education) was conceived as an attempt to institute change in mathematics instruction using such strategies. It was developed by the authors and teacher leaders at participating schools over a two-year period (1983–1985) and was in operation for four years (1986–1990). Funding was received from the National Science Foundation, the PL 98-377 State Grant Program, the University of California at

---

The authors were supported in part by National Science Foundation Grant TEI-8550283.

Santa Barbara, and the school districts. Schools in the project received external funding to participate for three years. The broad goal of the project was to improve mathematics instruction in participating schools by:

(a) Developing teacher leadership for school-based change in mathematics instruction;
(b) Assisting teachers to bring their instructional practices more in accord with constructivist[1] epistemology;
(c) Increasing teachers' knowledge about mathematics and changing their attitudes about mathematics and mathematics teaching;
(d) Improving collegiality among teachers;
(e) Raising educators' awareness of equity issues and changing teachers' practices with regard to race and gender equity;
(f) Changing assessment practices so that they reflect instructional goals.

Funding provided for a part-time Principal Investigator (Weissglass, a university mathematics professor), a full-time Project Director (Mumme, a former secondary mathematics teacher), teacher leaders (called teaching specialists in our project), release time for teachers, clerical support, and equipment. The schools in the project varied in size from 400–800 students in the elementary schools and 600–2400 students in the secondary schools. Overall, 65% of the students were white, 28% Latino, and the remaining 7% Black or Asian. The schools varied widely in the percentage of minority students—one school having a 68% Latino student population and another 3%. Schools were in rural, suburban, and urban settings, and students came from a variety of socio-economic backgrounds (e.g. middle class, working class, migrant farm workers).

The project had a profound effect on mathematics classrooms. An elementary teacher summed it up:

> By far the most lasting experience for me will be the experience of delight, by the children—for math. Math is loved by the children (by me too—and this has never been). Do you know they will stay in from P.E. to do math? They ask for hands-on lessons. Wow! For myself, I've gained confidence in myself as a math teacher. I'm not sure I would have said I was a math teacher three years ago.

A principal said:

> I see children thinking more and not just trying to get right answers. They are engaged in all math strands and are enjoying it. ... I see students relating math to the real world.

---

[1] Constructivism assumes that learning is the result of learners constructing their understanding through interaction with the environment. See Cobb (1986) for a concise summary.

Secondary teachers reported that as a result of Project TIME many changes had occurred at their school. Some of the things they indicated were: increased use of cooperative learning and manipulatives, reduced dependence on textbooks, more student enjoyment of mathematics, improved student/teacher relations, and increased collaboration among colleagues.

This article will describe (i) the project's view of mathematics instruction, (ii) the rationale for the project's approach, (iii) the intervention activities, (iv) the results, and (v) our conclusions and unresolved issues.

## I. View of Mathematics and Mathematics Instruction

We view mathematics as a way of looking at and making sense of the world. It is a beautiful, creative, and useful human endeavor that is both a way of knowing and a way of thinking. As a way of knowing, it consists of concepts and procedures constructed from empirical investigations and reasoning. As a way of thinking and reasoning, it involves exploring, investigating, conjecturing, evaluating, communicating, specializing, generalizing, abstracting, and justifying. These processes are as much a part of the mathematics we wish students to learn as any set of concepts and skills. It is through these processes that students expand their understanding of mathematical ideas and concepts. This view is different from seeing mathematics as a collection of memorized rules and procedures.

We believe that mathematical knowledge is the result of the learner's interpreting and organizing the information gained from their experiences. Developing students' mathematical thinking requires that teachers pay as much attention to *how* they experience the mathematics as to *what* mathematics they experience. The way students experience mathematics affects their perception of what mathematics is, as well as how well they understand it. We want students to experience mathematics in the classroom through projects, investigations, and small group activities. Investigations with concrete materials or situations provide the initial source of mathematical ideas. Even at the secondary level, abstract ideas find their source in concrete contexts. Periods of uncertainty and mistakes need to be accepted by teachers and students as part of the natural learning process.

Because students will try to relate new learning to what is already known (reconcile discrepancies, refine their thinking, and establish new connections), students should be provided with a variety of "opportunities to explore and confront any mathematical idea many times" (California State Department of Education (1987)). If students are allowed to construct their own understanding, they will come to see mathematics as a powerful and useful tool in their lives.

Students flourish in an environment that respects their thinking and feelings—where all questions and answers are treated with respect. It is the teacher's role to foster a questioning attitude in students. When students are encouraged to verify their own thinking, they do not need to rely on an outside authority to tell them whether they are right or wrong. Communication in the form of talking, listening, and writing, often through small group interactions, helps students to formulate and refine their thinking. (See Weissglass (1989) for a description of using small groups in mathematics instruction and Weissglass, Mumme, and Cronin (1990) for a summary of the effects of fostering communication in primary math classrooms.)

## II. THE REASONS FOR A SCHOOL-BASED APPROACH WITH ON-SITE LEADERSHIP

The view of mathematics instruction described above runs counter to longstanding instructional practices that are determined by teachers' beliefs and attitudes about schools, mathematics, and learning. We contend that changing these practices requires teachers to examine their beliefs about learners and learning, the nature of mathematics, and their role as educational decision-makers. Relationships among teachers and between teachers and administrators need to be reexamined. Issues of gender, race, and economic class must be addressed. Teachers and administrators need to experience for themselves that mathematics is a dynamic, exciting, and creative subject. They need to take risks, deal with their feelings about change, and construct new understandings of instruction. In addition, parents need to understand the changes in mathematics instruction, so they can be supportive of a school's reform efforts and interact mathematically with their children in productive ways.

The process of change is complex, and the path it takes is determined to a large extent by the culture existing in a particular school. Therefore, our strategy assumes that the proper focus of educational change efforts is the school as a whole. As Goodlad states, "There is little point in concluding that our schools are in trouble and then focusing for improvement only on teachers or principals or the curriculum. All of these and more are involved. Consequently, efforts as improvement must encompass the school as a system of interacting parts, affecting the others" (Goodlad (1984)).

It is not sufficient, however, to decide on a school-based strategy and then use traditional methods of professional development at the school site. Most past attempts at educational change have been attempts to change *what* (the information) teachers present. More recently there have been attempts to change *how* teachers teach—for example, using small cooperative groups, individualized instruction, or certain forms of lesson design. We believe it is necessary to address that *what* and *how* in relation to the *who*. In order to change mathematics instruction, it is necessary for teachers to change at a deeper level: how they relate to children; how they perceive themselves as teachers; how they feel about mathematics; how they relate to colleagues and administrators; and how they think about learning and schools. These more profound changes, if achieved, will certainly affect *what* and *how* mathematics is taught. The changes in the "what and how," however, will come from teachers' decisions and not because some authority is commanding it.

Traditionally, staff development activities at schools are led by site administrators, district personnel, and/or outside consultants. We believe, however, that these educators cannot provide the support necessary for the processes of personal growth. Although principals are essential to the change process, their responsibilities for management and evaluation make it difficult for them to provide the personal support necessary for change or to devote adequate time to lead reform efforts in any one curricular area. District level administrators can provide the atmosphere and validation so important for change to occur. They do not, however, have the personal relationships or the time necessary to support change at the site level. Outside experts can provide useful information, inspiration, support, and perspective, but they lack the understanding of the unique needs of the school. On-site leaders, to a much larger extent than outside consultants, are capable of transforming and being transformed by the reality of a school.

The project, therefore, funded a *teaching specialist* at each school—a teacher who was released from all classroom responsibilities to become an agent for change. It was through the teaching specialists' efforts that the project could effectively reach and respond to a large number of teachers. The teaching specialist brought to the change process an insider's knowledge of the culture of the school, the credibility of being a classroom teacher, ongoing relationships with the teachers and principals, a connection to the university, a long-term commitment, and a continual presence that would remain after project funding ends. This combination of assets is impossible for others to match.

Establishing the roles and selecting extraordinary teachers to fill them, however, was not sufficient. In order to deal with the complexity of the change process and to help teachers with their feelings about change, the teaching specialist needed both intellectual and emotional support. The authors' role was to provide that support.

In order to insure that beliefs, feelings, and attitudes were addressed by everyone involved in the project, we followed a model of educational change that gives equal importance to four components: obtaining information, reflecting and planning, acting, and obtaining emotional support. See Weissglass (1991) for a more complete description of the model.

**Information.** We wanted teachers to learn more about mathematics, teaching strategies, the learning process, and current issues in mathematics education. Engaging teachers in exciting hands-on learning activities increased their mathematical understanding, improved their attitude toward mathematics, and provided opportunities for them to discuss and reflect on pedagogy. Additional sources of information included articles from professional and research journals, curriculum development projects, and each other.

**Reflection and planning.** We believe that just as students must construct their understanding of mathematics, teachers need to construct (or reconstruct) their understanding of the teaching of mathematics. Therefore, teachers had extended opportunities to think about how they were currently teaching and what the alternatives were. They reflected, talked, and wrote about their assumptions, beliefs and reactions. They examined how their classroom practices corresponded to their beliefs. They wrote in journals and talked in pairs (called dyads) and in support groups,[2] as well as in discussion groups and informally. Issues ranged from the principles of constructivism to racial and gender bias, from the nature of intelligence to tracking. Teachers set goals about how they could be more helpful to students. They discussed how change in mathematics instruction relates to instruction in other disciplines. They planned how to implement what they were learning from the project in their classroom.

**Emotional support for change.** Although many educators are aware that teachers' feelings are a major obstacle to implementing change, very few attempts have been made to develop methods for dealing with this obstacle. Because feelings are difficult for many people to deal with, educational reformers often deceive themselves about this issue. They profess that feelings are not important, claim that their approach is adequately dealing with teachers' feelings, or ignore feelings completely. Project TIME took seriously Fullan's observation that, "Change is a difficult personal and social process of unlearning old ways and learning new ones. Deeper meaning and solid change must be borne over time. One must struggle through ambivalence before one is sure of oneself, that the new version is workable and right, or unworkable and wrong and should be rejected" (Fullan (1982)). Therefore, we assumed some responsibility to help teachers with their feelings about the proposed changes in mathematics education, as well as with their feelings about mathematics resulting from their own education. We employed methods to help educators learn how to listen better and develop emotionally supportive relationships. Our goal was for these enriched relationships to provide the support for teachers to work through feelings that inhibit their construction of new meanings and development of new approaches. We provide a variety of opportunities (support groups, discussion groups, one-to-one support, dyads) for teachers to talk and express their feelings about what was going on in their classrooms and schools and to construct new meanings.

**Acting.** Teachers tried out new mathematics activities in their classrooms. They set up learning situations for students to work in small cooperative groups, often using manipulatives to solve problems or initiate an investigation. They also participated in team teaching situations with the teaching

---

[2] Our use of dyads and support groups gives people the opportunity to be listened to without their thoughts and feelings being analyzed, interpreted, or argued with. See Weissglass (1990).

specialist, conducted Family Math evenings for parents, and made changes in the ways that they related to colleagues and administrators.

## III. PROJECT ACTIVITIES

Although the project's ultimate goal was to affect children's mathematical experiences, the project activities directly addressed four groups: teaching specialists, teachers, principals, and parents.

**Teaching Specialists.** Each teaching specialist is a classroom teacher who had participated in previous leadership development activities with the Tri-County Mathematics Project, a site of the California Mathematics Project. The Tri-County Mathematics Project runs a four-week summer institute each summer for twenty-five to thirty teachers grades K–12. Each teaching specialist attended one summer institute and participated in the leadership of a second. As part of that second institute, the teaching specialist participated in a four-day leadership seminar which preceded the institute.

Project TIME released the teaching specialist from all regular classroom duties. Her or his role was to think of ways to help teachers to improve their mathematics teaching and provide thoughtful intervention. The teaching specialist listened to and supported teachers and the principal; helped teachers develop and stand up for school policy; led professional development activities; and acted as a communication link between teachers and district office and between the school and the university.

Each week the teaching specialists and the authors participated in a day-long seminar. At these seminars we listened to the teaching specialists and provided a supportive structure for them to think about how to solve their site-specific problems. The seminars provided the opportunity for teaching specialists and staff to have dyads, share ideas, learn more about mathematics and mathematics education, reflect on the change process, and plan school and project activities. A typical seminar would start with people sharing, briefly, successes from the previous week. Then there was either a support group or a long dyad (fifteen or twenty minutes for each person) for people to reflect on and express feelings about what was going on in their work. (The authors participated as peers in the support group and dyads.) Either a math activity followed by a discussion or a discussion of an article about mathematics education would follow. There would also be time for teaching specialists to plan together in small groups for the professional development activities they were leading. Sometimes the whole group would plan for a project-wide activity. The seminar concluded with a closing circle where each person had a chance to express what he or she got out of the day and anything he or she would like to see addressed at the next meeting. Twice each year staff and teaching specialists participated in a two-day retreat.

The teaching specialists planned and concluded professional development activities specific to the needs of their site. These included all day

> **Secondary School Inservice Agenda**
>
> 8:30  Coffee and informal discussion
> 8:45  Support Group—Share successes and frustrations
> 9:45  Break
> 10:00 Math Activity—Sampling
> 11:15 Discussion of activity and implications
> 12:00 Lunch
> 1:00  Sharing what you've done in your classroom
> 2:00  Planning—How can I use this in my class?
> 2:45  Evaluation and what next?

FIGURE 1

workshops (eight the first year, six in subsequent years), demonstration lessons in teachers' classes, team planning and teaching, and peer coaching. They also organized and led support groups for teachers and met individually with teachers to listen and to offer support and encouragement.

Professional development seminars included discussion groups and time for teachers to pair up and listen to each other's thoughts or feelings. For example, after a hands-on mathematics activity teachers would discuss one or more of the following: "How did this mathematics experience contrast with how you learned the subject matter in school?" "How did you feel about working in a group?" "Was there anything the group could have done to have been more helpful to you as a learner?" "What would be the challenges of implementing such activities in your classroom?" "How would you feel if mathematics at your school were taught this way?" They would also address questions such as: What are your assumptions about learning? Remember some learning experiences that occurred outside school. How were they different from schooling? How were they similar? When did you have to memorize information? How did you feel? Why did you do it? How much of the information do you still remember? Are you able to use it? An agenda for a typical workshop is in Figure 1.

**Teachers.** Participation in the project was voluntary. At each site a 70% affirmative vote of the faculty on an anonymous ballot was required before they were admitted to the project.

Project activities began with a four-day residential workshop for all schools. Each subsequent year began with a two-day summer workshop. There were eight all-day workshops at each school during the first year and six during the second and third years. Workshops included small group hands-on mathematics activities, discussion of issues, review of the research, time for reflection, and opportunities to engage in planning with other teachers.

An example of an activity[3] that we used with teachers and which they adapted for their students is outlined in Figure 2 (see p. 12). It illustrates one way probability and statistics are used in the real world to predict diseases from blood samples. It shows the applicability of mathematics in the world, involves a variety of mathematical processes, and can be extended to show the usefulness of computers in simulating real world situations.

The histograms in Figure 3 (see p. 13) are shown to the teachers. Each box is labeled with one name from the council and has ten cubes distributed according to one of the histograms. The person assigned Disease X has two blue, five red, and three yellow cubes, but that histogram is not shown.

The questions we asked teachers to reflect on and discuss after the activity were:

(1) What and how did you record information and how did this help you?
(2) How did you decide when to stop sampling?
(3) Would the results be the same if there were 100 cubes in the box?
(4) In what ways were you encouraged to develop your own methods and verify your own thinking?
(5) How did this experience differ from the way you learned mathematics in school?
(6) What view of mathematics does this communicate to students?

Teachers were expected to participate in support groups for approximately four hours per month. Each school decided the best time to do this. Some faculties met during lunch and some met after school. In one school, the faculty divided into two groups and met for two mornings each week from 8:00–9:00 a.m. Since the students arrived at 8:25, the principal conducted an assembly for one-half of the student body until 9:00 a.m. in order to release the teachers for support group. During support group meetings teachers expressed their feelings about change, reflected on a variety of educational issues, gave and received support, and learned how to listen better.

Teachers observed demonstration lessons given by teaching specialists in their classrooms. The teacher and teaching specialist planned and taught lessons as a team. This practive evolved, in some cases, into a coaching relationship. Teachers were also expected to try out new ideas in their classroom—the time and content determined by the teacher. The following passage from one third grade teacher gives some indication of the process teachers underwent in their classrooms, their thoughtfulness about the teaching/learning process, and the way they struggled with the ideas promoted by the project:

> The biggest change in me is that I feel less like a teacher and more like a researcher, or maybe, I'm getting the idea of what

---

[3]The activity was adapted from an idea of William Finzer. A related computer simulation is contained in Finzer et al. (1984).

### Sampling on the island of *Xanadu*

**Grouping:** Groups of 3-4

**Materials:** Prepared boxes with cubes to simulate taking "blood samples."
(A corner is cut away from the box so one cube can be seen at a time.)

**Goals for students:** Drawing inferences and predicting from samples
Simulating an application from the real world
Working together cooperatively
Representating data graphically

(Students should have had some previous experience with elementary probability and representing data graphically.)

**Background Information:** This activity simulates one way probability and statistics are used in the real world—predicting diseases from blood samples. Each group will be a team of medical researchers who have been flown to the small, isolated island of Xanadu to solve a medical crisis. There is an epidemic of a new disease which is caused by overexposure to video games. One of the members of the ruling council has the disease. Groups are to test the blood of members of the ruling council (Theano, Wonder Woman, Batman, George Bush, and Dick Tracy) to determine what each person has and who has the dreaded Disease X. No one knows what the blood sample looks like for disease X, but they do know what the blood samples look like (See Figure 3.) for the following conditions: normal, mono, allergies, and strep. We know that each person has one of the five conditions, more than one person can have any of the conditions, but only one person has Disease X.

**Procedure:** Each team receives boxes representing blood from each of the five members of the council. Teams simulate taking blood sample by repeatedly shaking the box, looking at one "cell" at a time, recording its color. They keep track of their sampling and are provided with graph paper to encourage graphing their results. The team members should discuss their results, compare with the histogram, and decide when they are fairly confident that they know which of the conditions the blood sample represents. If Disease X is not correctly diagnosed, the patient dies!

Medical technology is very expensive on Xanadu, since consultants' salaries must be paid, so each "cell" sampled costs $1,000. The entire Xanadu budget is only $150,000, and money not used for blood testing will go to the local schools.

Teams post results on a class chart and discuss results. The object is to make the most accurate diagnoses and spend the least money. For diagnoses that are in question, have participants graph cumulative results on poster-size graph paper.

FIGURE 2

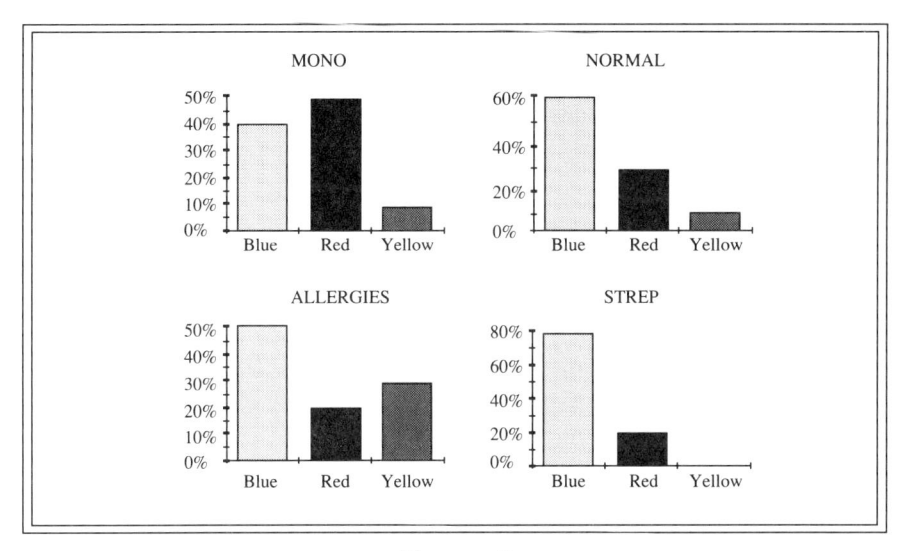

FIGURE 3

Teacher really is. ... I'm delighted with how the kids love math this year. We've not opened a book for months and first they thought we were just fooling around with tiles. ... Then as we continued, mainly working on developing the concept of multiplication and discovering patterns of multiples, they (most) caught on to the exploration. ... I'm amazed how some kids just don't get it. Day after day we create these things and some can't make the transfer from tiles to beans, or see how the tiles relate to a word problem. A number still don't see how a grid pattern is the same as an algorithm (how $5 \times 4 = 20$ is the same as 5 tiles down and 4 across). I get the sense that tiles are as much an algorithm to them as the symbolic numbers ....

However, the kinds of ahas! are of a more sophisticated level, and most of the kids do really get it, I think. ... Every one of my kids can write addition and subtraction word problems. Many of them still can't create a multiplication situation and almost none can write a division situation after all the manipulative work we've done and the verbal situations we keep inventing. I wonder if it's too complicated. If it is, maybe we need to postpone multiplication and fractions and certainly decimals for a year or more and just play with patterns and measurement and maybe some logic after 3rd grade. ... Plenty of other days I'm delighted with the whole experimental process. Wonderful spontaneous things are happening. Kids interact creatively... more than before.

Empowering teachers to initiate and sustain personal and school strategies to achieve equity for females and minorities was an important goal of the project. We raised teacher awareness of the effects of current instructional practices on minorities and females through activities (such as those from EQUALS[4]), videotapes, discussion of articles, and having teachers who were members of minority groups tell about their own school experiences.

**Principals**. Principals attended five half-day seminars with their teaching specialists during the academic year. The purpose was to provide for communication between administrators at different sites and between the authors and principals. The seminars usually started with teaching specialists and principals participating in a mathematics activity and discussing the implications of that activity for instruction and staff development. This was followed by groups discussing the implementation of the project at their sites. Successes were related and problems discussed. At each seminar the principals had a chance to meet as a support group with one or both of the authors.

The principal also participated in workshop activities with his or her teachers as much as possible. Principals encouraged and supported teachers who were taking risks with new ideas, and the principal and teaching specialist supported one another in their mutual efforts to improve mathematics instruction.

**Parents**. Teaching specialists and teachers conducted parent education sessions, including Family Math (Stenmark, Thompson, and Cossey (1986)) and produced newsletters at each school. Teachers and Teaching Specialists made presentations at school board meetings, and the project conducted a workshop for school board members and parent leaders. Parent surveys indicated that they felt that mathematics instruction had improved. After attending parent education sessions one parent wrote:

> Each of the Family Math evenings has been delightful—partylike, playful—yet mathematically serious. ... We are convinced that all of the children at our school will learn more math, enjoy math more, and have a better development of their general analytical skills as a consequence of your dedication to this program.

## IV. THE RESULTS

The outcomes of the project were monitored by surveys given to participating teachers and parents. Teachers' written opinions were periodically solicited. In addition, teacher leaders and principals were interviewed by an outside evaluator.

Change occurred in several areas: classroom practices, teacher attitudes, communication, and district practices. We include in this description, mainly, written responses from the teachers. The numerical data presented is done informally. Any statistical analysis would be affected by the fact that 27% of

---

[4] EQUALS (Lawrence Hall of Science, University of California, Berkeley) develops programs and materials to address issues of equity in mathematics and science education K–12.

teachers responding to the postsurvey were teachers that did not respond to the presurvey, having joined the school either in the second year (15%) or or third year (12%) of the project.

**Classroom practices**. The average number of hours per week of math instruction at the four elementary schools[5], after two and one-half years in the project, increased 20% from 4.2 hours to 5.1 hours. Furthermore, teachers reported that they increased the breadth of the mathematical experiences for their students by decreasing instructional time devoted to number and increasing the time for other areas. The changes are indicated in Figure 4.

The survey also showed that the percent of class time that students used manipulatives increased from 28% to 35%, and the number of teachers using various manipulatives increased substantially. In addition, the percent of teachers using calculators increased dramatically from 23% to 74%.

Teachers reported improvement in student attitudes toward mathematics. In addition, some secondary teachers reported improved attendance in classes where attendance was traditionally a problem. They attributed the improvement to the incorporation of more relevant hands-on activities.

Evaluation practices also changed. Teachers were asked what weight they assigned various modes of evaluating or grading students. The data, as shown in Figure 5 (see p. 16), indicates a decrease in traditional modes.

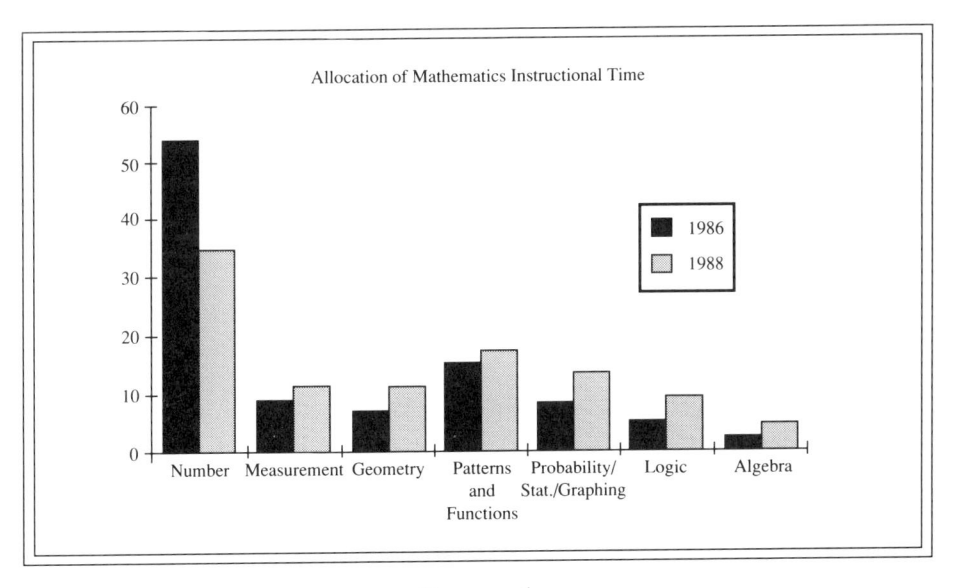

FIGURE 4

---

[5]Numerical data is presented only for four elementary schools that completed three years of the project in the first cycle (one to three years) of the operation. Seventy-four teachers completed the final survey. The number of teachers in the secondary schools in the first cycle is too small, and the turnover was too great to provide meaningful data. One elementary school and three secondary schools completed three years during the second cycle (years 2–4), but there was insufficient funding to evaluate the results from these schools.

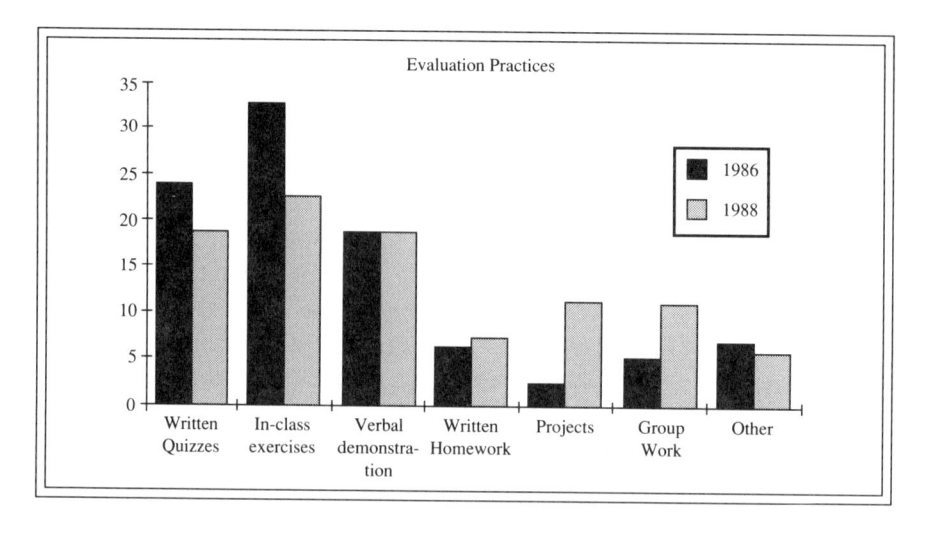

FIGURE 5

Several teachers participated in an Assessment Interest Group which met regularly for two years. They discussed assessment alternatives and tried new ideas in their classrooms. One teacher commented that working in the assessment group, "was very helpful. It caused me to re-examine my notion of understanding and gave me the confidence to probe deeper."

Another commented:

> I realize now that there aren't any clear formulas for judging of our children. ... It has clarified the importance of using many ways to assess and that each child may need something different.

**Teacher attitudes.** The project achieved success in improving teachers' self-confidence. One teacher wrote:

> I actually feel myself coming to a place where I feel safe in my own learning process! I don't have to know it all to feel confident in teaching or to share what I know with others. Now I want that same feeling for children! I want them to feel safe, confident and able to share what they know without fear of what they don't know. I hope they can unlock those doors and communicate understanding while enjoying the awesome, wonderful experience of learning mathematics.

And another:

> I found that there are things that I can do as an educator to improve gender and racial differences within our society.

Teachers became more aware of significant issues in mathematics education. They wrote:

> These questions constantly recur: how am I influencing my students and am I constantly aware of my influence? My awareness has been heightened as a result of the discussions of equity.

> I realize that standardized tests and publisher-made tests are not true tools for assessment, but I don't really know what to use in their place.

**Mathematical understanding.** We did not conduct a direct assessment of teachers' mathematical knowledge because we felt this would place undue pressure on them. There is, however, considerable evidence from our observations of teachers in seminars that their understanding of mathematics and ability to solve mathematical problems increased. There are also the teachers' own reports. A typical comment was:

> I have a new confidence in my own mathematical/problem solving ability which stimulates me to use more math in all subject areas with my own students. It's wonderful to believe now that mathematical learning blocks can be overcome.

**Communication.** Teachers talk more to one another and have established collegial working environments at project schools. Typical teacher comments are:

> I see all of us reaching out more. It is a good feeling knowing you have someone who will listen and offer a helping hand.

> I find that the support groups are very helpful in venting difficulties or concerns. I think they are also a good way to get to know my colleagues and feel like I can go to them with ideas or questions. Teaching in isolation is very frightening.

Support groups have sometimes been difficult to implement, however, and some teachers have not liked them. The prevailing culture in many schools results in teachers working in isolation. Schools are not usually a place where teachers share feelings with each other, except possibly to gripe or complain. Some teachers felt that support groups added one more burden to an already overcommitted schedule. Within some faculties there were existing groups of people who disliked one another. This made it difficult to establish the trust necessary for support groups to function well. In addition, some teachers disliked the structures and rules established for support group operation. It is interesting to note however that in all schools, even those where support groups experienced problems, teaching specialists and principals report increased collegiality and morale.

**District practices.** All of the districts have reformed their staff development practices to some extent and have utilized the leadership developed by the project. Two districts have instituted lead teacher programs (scaled down versions of Project TIME). One small, unified district employed a secondary school teaching specialist as a half-time math specialist for the district after project funding ended. This district also employed an elementary teaching specialist as a curriculum resource teacher at her school. In the secondary schools our model for staff development has spread to other departments and teaching specialists have done in-services with other secondary math departments in their district. One high school has implemented a major school restructuring effort, a "school within a school" growing out of and involving many of the ideas from Project TIME. Several of the former teaching specialists are now mentor teachers.

## V. Conclusions and Unresolved Issues

**Change is slow.** Teachers need time to comprehend the proposed changes in mathematics instruction. Most teachers were taught mathematics by rote in competitive teacher-centered classrooms. Valuing mathematical processes, rather than memorizing facts and algorithms, and promoting cooperative, student-centered learning experiences is new to them. Once they understand this view of mathematics learning, they then have to integrate it with their own teaching. In three years teachers only begin to transform what happens in their classrooms. Secondary teachers are slower to adapt constructivist approaches. Five or ten years is a more realistic time frame for achieving fundamental reform; and, even then, there will be new changes to face. Three years is enough, however, to institute the processes that will lead to sustainable change.

**Change is complex.** Although we recognized when we started that change was complicated, we now know that it is far more complex than we initially imagined. Many things at the school district interact to affect mathematics education. Therefore, everyone, including board members and district administrators, must have some understanding of the proposed changes. Having "the blessing" of the district administration is not enough. Time and resources must be undertaken to gain their understanding, not only of the nature of mathematics, but also the nature of the change process.

**Change is possible.** We have come to regard change as an ongoing process. The project processes of learning through hands-on experiences, reflecting, dealing with feelings in dyads and support groups, and implementing new ideas became a regular part of many teachers' professional lives over the three years. Whether districts and schools will provide enough structural and financial support for these processes to be sustained is still to be seen. The fact that principals from some schools no longer funded by the project continue to meet with project staff to discuss these issues is encouraging—the frustrations they feel about getting district support is not.

**Change is costly.** There are no quick fixes or easy answers to providing the resources necessary to support change. These resources are expensive. The primary change agent, the teaching specialist, is released from all regular classroom duties to provide full-time support for the teachers at school. The teaching specialists also require support. It is not sufficient to create the role of teaching specialist. A coordinator is needed to bring teaching specialists together to network and provide mutual support. Significant periods of quality time are required for teachers to engage in critical inquiry. Teachers need adequate resource materials, manipulatives, and appropriate classroom equipment to implement new ideas.

**Change is facilitated by supportive relationships.** It is our belief that the ability of the teaching specialists to listen to teachers and the establishment of structures for the teachers to listen to each other and work through their feelings about change increased teachers ability to change their mathematics instruction and to consider emotion-provoking issues in education. We do not have any hard data on this. It would be necessary to do a long-term study of social relationships at a school and how that affects instruction. Certainly the structures established to allow the teaching specialists to express their feelings were crucial in enabling them to make the transition from classroom teacher to teaching specialist and to function effectively in that role. The teaching specialists are unanimous on that point.

**Change efforts in mathematics education affects more than mathematics instruction.** We have found, especially at the elementary schools, that ideas learned in the context of mathematics education were applied to all areas. For example, small cooperative group approaches were used in other subject areas, sometimes even before they were used in mathematics. The process of observing and discussing children's mathematical thinking spread to other curricular areas and led to increased respect for children's thinking. At the secondary level, other departments adopted project methods in their staff development efforts.

In all schools the entire climate was improved by increased attention given to thinking about children and the learning process, dealing with teachers feelings, and breaking down the isolation between teachers.

**Change in mathematics instruction is hindered by other factors.** Some of the most important factors that adversely affected our ability to bring about change are listed below.

*Testing.* Proficiency tests, standardized tests, and other commercially prepared tests all affect teacher and student perceptions of what is valued in the curriculum.

*Materials.* The dependence on textbooks and the unavailability of adequate alternative curriculum materials limits teachers' abilities to implement changes in the mathematics classroom.

*"Continua."* A sequential listing of topics and skills often inhibits teachers' abilities to think about mathematics as an integrated whole or allow students to experience its full breadth.

*University entrance requirements.* A perceived need to "cover" the curriculum to meet university requirements creates pressure in secondary schools to maintain the status quo.

*Behavioristic teaching models.* Wide-spread models of teaching based on behavioristic principles dominate many school environments. The inconsistencies between the project's approach and the models are not always readily apparent to the promoters or practioners of behavioristic approaches. Yet assumptions about learners and learning, which exist at the very core of behavioristic models, are contradictory to the goal of empowering learners.

*Models of classroom discipline.* Empowering students in mathematics requires creating a total learning environment that respects the learner and builds upon intrinsic motivation for learning. Popular models of classroom discipline that are based on behavioristic reward and punishment practices interfere with developing a constructivist approach in the learning environment.

*Tracking (ability grouping).* Children educated in a tracked system receive the message that some children are less capable than others. It discourages all children from reaching their full potential as learners and discriminates against children from minority and low-income families.

*Educators' attitudes.* Teachers', administrators', and school boards' willingness to change mathematics instruction is affected by sexism, racism, and attitudes about learning; the role of schools in the society; and the need for change.

***

The changes achieved by Project TIME are unlikely to be sustained unless some of the above factors are changed. When teaching specialists return to the classroom, they encounter the typical difficulties of again being a classroom teacher compounded with the stress of a change in roles. They do not have the resources to continue their leadership role as an extra responsibility. Although some districts have provided some released time for them to provide leadership, they still are under stress. We did not think through clearly enough the issues related to the end of funding. It would have been helpful to require, from the beginning, that districts regard the project as starting a process that needed to be incorporated in some way into "business as usual" after funding ended. If schools are already considering "restructuring," approaches similar to ours could be used to good advantage to provide the curricular focus for reform efforts. However, for such reforms to result in changes in classroom practices that are sustainable, district administrators and school board members, as well as teachers, need to reassess their assumptions and priorities. If districts continue to assess school performance by means of results on standardized tests, then they will not support teachers in their change efforts. Schools' change efforts will wither in an unsupportive district.

Although it is easy to be skeptical, there is some reason for optimism. Mathematics educators are expending considerable effort toward developing policies (National Council of Teachers of Mathematics (1989), National Research Council (1989)) that incorporate elements of a constructivist approach to mathematics learning. Curricular materials are being developed to implement this philosophy. There is increasing resistance to behavioristic models for teaching (Confrey (1986)) and for classroom discipline (Gartrell (1987)), as well as to standardized testing (National Association for the Education of Young Children (1988), Neil and Medina (1989)). The negative effects of tracking are being documented (Oakes (1985)). There is increasing awareness of the importance of addressing educators attitudes and beliefs in change efforts (Fullan (1982), Hall and Hall (1988), Hall and Hord (1987), Levine (1989), Ost (1989), Weissglass (1990)).

At a project-wide workshop concluding our three-year effort with a group of schools, one principal reported from a discussion group:

> The good news is that change has occurred. The bad news is that we know what it requires to achieve it.

We started off our project believing that there was no quick fix to improving mathematics instruction in our schools, and we are even more convinced of that today. One may wish that there were simple answers—that a curricular package, an assessment package, or a teaching skills package would bring about change in instruction, but real change is messy. It involves triumphs and frustrations, emotions, and turmoil. An old Chinese proverb says, "The trees may prefer calm, but the wind will not subside." We may wish the change process to be calm and predictable, but even in an area (mathematics) that is (incorrectly) assumed to be devoid of emotion, it is not that way.

With this cautionary note we encourage others to take what we have learned and adapt it in new ways so that our schools will be better places for young people to learn and grow.

## REFERENCES

California State Department of Education (1987), *Mathematics model curriculum guide, kindergarten through grade eight*, California State Department of Education, Sacramento, CA.

Carnegie Task Force on Teaching as a Profession (1986), *A nation prepared: teachers for the 21st century*, Carnegie Forum on Education and the Economy, New York, NY.

P. Cobb (1986), *Making mathematics: children's learning and the constructivist tradition*, Harvard Educational Review **56** (3), 301–306.

J. Confrey (1986), *A critique of teacher effectiveness research in mathematics education*, Journal for Research in Mathematics Education **17** (5), 347–360.

J. A. Dossey, I. V. S. Mullis, M. M. Lindquist, and D. L. Chambers (1988), *The mathematics report card*, Educational Testing Service, Princeton, NJ.

W. F. Finzer, J. Gutierrez, and D. Resek (1984), *Math worlds: exploring mathematics with computers*, Sterling Swift, Austin, TX.

M. Fullan (1982), *The meaning of educational change*, Teachers College Press, New York, NY.

D. Gartrell (1987), *Assertive discipline: unhealthy for children and other living things*, Young Children **42** (2), 10–11.

J. I. Goodlad (1984) *A place called school*, McGraw-Hill, New York, NY.

E. Hall and C. Hall (1988), *Human relations in education*, Routledge, London, UK.

G. E. Hall and S. M. Hord (1987), *Change in schools*, State University of New York Press, Albany, NY.

S. L. Levine (1989), *Promoting adult growth in schools: the promise of professional development*, Allyn and Bacon, Boston, MA.

National Council of Teachers of Mathematics (1989), *Curriculum and evaluation standards for school mathematics*, NCTM, Reston, VA.

National Research Council (1989), *Everybody counts: a report to the nation about the future of mathematics education*, National Academy Press, Washington, DC.

National Science Board (1983), *Educating Americans for the 21st century*, National Science Board, Washington, DC.

D. M. Neill and N. J. Medina (1989), *Standardized testing: harmful to educational health*, Phi Delta Kappan **70** (9), 688–697.

J. Oakes (1985), *Keeping track: how schools structure inequality*, Yale University Press, New Haven, CT.

D. H. Ost (1989), *The culture of teaching: stability and change*, The Educational Forum **53** (2), 163–181.

J. K. Stenmark, V. Thompson, and R. Cossey (1986), *Family math*, Lawrence Hall of Science, Berkeley, CA.

J. Weissglass (1989) *Cooperative learning using a small group laboratory approach*, Cooperative Learning in Mathematics, A Handbook for Teachers (N. Davidson, ed.), Addison-Wesley, Reading, MA.

―― (1990), *Constructivist listening, for empowerment and change*, The Educational Forum **54** (4), 351–370.

_____ (1991), *Teachers have feelings: what can we do about it?* Journal of Staff Development **12** (1), 28–33.

J. Weissglass, J. Mumme, and B. Cronin (1990), *Fostering mathematical communication: helping teachers help students*, International Perspectives on Transforming Early Childhood Mathematics Education (L. Steffe and Wood, eds.), Lawrence Erlbaum Associates, Hillsdale, NJ, pp. 271–281.

CENTER FOR IMPROVING MATHEMATICS AND SCIENCE EDUCATION, UNIVERSITY OF CALIFORNIA, SANTA BARBARA, CALIFORNIA 93106

# Diagnostic Testing: One Link between University and High School Mathematics

ALFRED MANASTER

## INTRODUCTION

Each fall at the University of California, San Diego (UCSD), approximately 1,800 students register for a beginning calculus course or a course to prepare them for calculus. Most are entering students who want to take calculus. Some simply need to satisfy a college breadth requirement, but many understand the need for calculus to prepare themselves for majors in the sciences, engineering, and such social sciences as economics. For many years, UCSD has had two calculus sequences—one directed toward the needs of students who will later study physical sciences or engineering and the other to serve those students intending to major in other fields. The extent to which students are ready to take calculus is not easily measured by the amount of their prior mathematics study since their mastery of that material varies more than their grades and since their understandings and skills may have faded with the passage of time since their last exposure to them. To assist students in selecting a first mathematics course, the Mathematics Department has offered a placement test for more than fifteen years.

In 1980, I was asked to direct the UCSD mathematics placement testing program. My responsibilities included both developing or selecting a suitable test and overseeing the administration of the program. At first, the placement testing program was to be used by the Department of Mathematics to assure that students had the prerequisite knowledge for the courses in which they were placed so that they would be likely to succeed in their first and subsequent college mathematics courses. Later it became clear that counseling or allowing students to register for courses for which they are overprepared is just as harmful as the more obvious error of placing students in courses for which they are underprepared. Overprepared students get bored, develop very poor study habits, and have great difficulty readjusting when new material is finally presented. Indeed, they may develop a false sense of security and be lulled into complacency which comes from working on a surface level while not coming to grips with the issues of the course.

The department was also interested in reviewing and revising its courses for the stronger mathematics students. If we had the confidence that students had the necessary mathematics background and facility, we could avoid spending excessive time on review and would not be frustrated by failing many students or compromising the level of the course.

I needed a test that would meet these criteria, i.e., finding students' deficiencies and their strengths. Fortunately for me, my attention was directed to a test that had recently been developed by the Mathematics Diagnostic Testing Project (MDTP) of the University of California and The California State University. The test was designed to determine students' readiness for college calculus. Initial field testing on several college campuses showed that students' test scores had a much higher correlation with first term calculus grades than did any other available measure. The correlation was approximately three times greater than any combination of high school grades, high school mathematics grades, and SAT mathematics scores. The reliability of the test coupled with its availability for my campus, appeared to solve my immediate problem. Because of my interest in the test, I also became a member of the workgroup for MDTP.

I did not know at that time that my participation in the workgroup would lead me to become involved with and challenged by new questions and fundamental issues affecting mathematics education. Why do our students so often disappoint us? What essentials are they missing? Do they fail to know mathematics because we do not cover the material in our classes, or do they fail to learn what we do teach? What mathematics should all college students master? What mathematics should all high school students master? What are the crucial aspects of mathematics for later success in mathematics, science, or life in an information society? How can we increase the chances that all students will reach their (substantial) potentials for reasoning carefully and thoroughly and will understand the mathematics they need? How can we share our enjoyment of mathematics with our students and communicate it to the public? How can we help to influence and form public policies that will support these goals? What are effective ways for high school and college mathematics faculty to work together to strengthen the true mathematical literacy of our society?

In this paper, I would like to describe how mathematicians, through their involvement in diagnostic testing projects such as the California Mathematics Diagnostic Testing Project, can gain a better understanding of such issues and how they can contribute to improving the conditions and content of mathematics instruction. My experience indicates that communication among mathematics faculty at many levels is enhanced through their participation in diagnostic or prognostic testing and that people join together to pursue their common goals.

## A Brief History of MDTP

The California State University and the University of California established MDTP in 1977. They charged the workgroup to determine the mathematics competencies necessary for success in college calculus, to develop a diagnostic test for those competencies, and to establish a communication network with California high schools for exchanging information about their students' mathematical preparedness for college mathematics and science courses. I joined the workgroup in 1980. Since its formation, the workgroup's members have developed a more complete understanding of its mission. To increase both the amount of mathematics that each student studies and the quality of that student's mathematical experience and work, we need to understand many components of mathematics preparation at all levels. As college and university mathematicians, we need to work with our mathematics colleagues in other segments of education and our policymakers in the educational community to assure that teachers and students have the supportive environments they need to teach effectively and to learn well.

The initial composition of the workgroup included mathematics and science faculty from each of California's public university systems, mathematics faculty from its community colleges and its high schools, and a staff member with experience in mathematics education. The composition has changed slightly with the addition of more high school and community college mathematics faculty. The perspectives of teachers of freshman calculus and science, high school mathematics, and community college preparatory and calculus courses are all of great value in understanding what needs to be tested and why. Experienced high school and community college teachers contribute a good sense of the kinds of errors students make. College mathematics faculty bring an understanding of the ways in which various topics and techniques are used in learning calculus. Science instructors provide a context in which calculus will be used as well as an insistence on using reasonable data and measurement units. This variety of perspectives helps to provide item distractors that attract students at various levels of preparation and with differing patterns of errors.

The workgroup first developed a set of topics (content based) that, in its members' judgment, were essential for success in college calculus. (A list of the topics with typical items for each, that is made available to students before taking the test, appears as an Appendix.) The workgroup then developed items to test specific skills and concepts in those topics. Most importantly, the first test was field-tested with great care. Students were tested at the start of various college calculus courses. Their performance on the test was then compared to their course grades. Not only were the overall test scores examined, but the topic scores and the responses to individual items were also. Both the class grades and the overall test scores of students selecting each offered response (option) to a question (item) were carefully reviewed.

Care was taken to ensure that incorrect options (distractors) which attracted good students were seldom offered. Not only were ambiguous distractors avoided, but those distractors that result from errors that good students tend to make were usually eliminated.

There were two factors that led to the widespread adoption of the tests developed by MDTP on campuses of California's public universities. One factor was the increasing perception of the need for effective counseling for entering freshmen based upon the best available information about their mathematics preparation. While the faculty was faced with increasing numbers of students who needed calculus but did not have adequate preparation to do well in it, there was also growing awareness of the importance of providing enough support to admitted students so that they could have academic success. The second factor was the effectiveness of the MDTP tests. Two indicators of that effectiveness were the strong correlation of test scores with future course performance and the face validity found by mathematics faculty who reviewed and then adopted the tests.

During the 1981–1982 academic year, the test was given to over 35,000 university students. By 1984–1985, MDTP tests were used on all eight general campuses of the University of California and eight of the nineteen campuses of The California State University. The decision to use the tests on a campus is always made by its mathematics faculty. In turn, the mathematics faculty's support gave the tests credibility when they were offered to high school teachers in the state.

While the calculus readiness test was being accepted for counseling and placement use on university campuses, some high school teachers who learned about it asked for copies to test their students. They wanted to see the extent to which their calculus preparatory students were meeting this measure of student readiness for calculus. The initial informal use of the test in the high schools led to a much wider program. It provided an invaluable setting for developing an effective reporting system and communication network.

In addition to refining and revising the calculus readiness test, the MDTP workgroup recognized a need for a test to indicate readiness for the college course that prepares students for calculus, often called mathematical analysis. The same test development process was used to create, field-test, and release that test. Similarly, a test of readiness for intermediate algebra was developed. Almost all its field testing was done on campuses of The California State University and in community colleges, since very few students take a course at this level in the University of California.

Simultaneously, increasing numbers of high school teachers were using the tests, providing suggestions for its dissemination, and requesting a test to determine levels of readiness for beginning algebra. In 1985 an algebra readiness test was developed. Because very few university students in California take a first-year algebra course, field testing was done in selected high schools. The two measures of course performance used to validate the test were taken near the end of the algebra course; they were teacher assessments of readiness for the next course and results on the MDTP test of readiness for second-year algebra. After subsequent revisions during the summer of 1986, the test was made available to high school and junior high school teachers in California. It was given to approximately 122,000 students during the 1986–1987 academic year.

Several structural changes were made in the project to allow for better responses to the increasing demand for tests in the early part of this decade. In 1981–1982 the project office at Berkeley distributed materials for testing 17,000 students to high school teachers. Two years later project centers on five university campuses distributed materials for testing 138,000 students and scored 67,000 tests for high school teachers in the state. In the 1987–1988 school year, project sites on ten university campuses scored tests for almost 284,000 junior and senior high school students in over 800 schools throughout California.

## High School Services of MDTP

Each project site distributes test materials and returns detailed test result reports to teachers. Test scoring is done promptly, usually within two working days. More importantly, each project site is a resource for teachers in its service area, serving as a natural link to the university campus. Information about the tests, their significance, and related curricular issues are provided through local user conferences, visits to schools and districts, and open informal communication. Almost all site directors teach mathematics; those sites directed by administrative staff have close working relations with their campus' mathematics departments.

Teacher requests account for almost all of the use of the tests in high schools. The typical, and clearly the most effective, pattern for introducing the tests in a school begins with one teacher who first uses one or two of the tests. That teacher then discusses the tests and the results with some colleagues and orders materials for them the next year. By the third or fourth year, almost all the teachers want to use the tests. We try to arrange for one teacher to order for the school. Results are sent back to each teacher, although the envelopes are often placed into a box sent to our school contact. When there is substantial use in a school, we offer to provide summaries to the department, but without disaggregating the summaries in ways that would invite class or teacher comparisons. Teachers at some schools analyze the results and review the analyses with their colleagues as part of their curriculum reviews and as part of their counseling recommendations.

Discussions with teachers have led to many enhancements of the services that MDTP offers—the most notable is providing a letter summarizing the test results to each student. When we learned that teachers were cutting apart their lists of students and topic scores to give each student a strip with his or her scores, we realized that teachers valued the diagnostic information we were trying to provide. Fortunately, adequate computer facilities were already available so that we could provide reports to each student, via the teacher. We have included this information in our service to schools since 1983.

We have also listened to teachers to find ways to provide information that will be easily accessible and useful to them. Providing graphical or other clear summaries is very helpful to many teachers who have very limited time to analyze test results. Teachers suggested including reports of how many items each student answered and the last item each student answered: these data provide helpful indications of some aspects of the student's performance, including reading skills and careful or careless work. As mentioned earlier, teacher requests also led to the development of the algebra readiness test.

## Uses of the Diagnostic Tests

The workgroup had considerable concern that the algebra readiness test might be used as a barrier to algebra rather than as a tool to help prepare students for algebra. Several steps were taken to discourage this abuse of the test. We emphasized the diagnostic nature of the test through newsletter articles, user conferences, visits by project personnel to schools and districts, and a special mailing to users. We encouraged administering the test rather early in the prealgebra course so that teachers would have indications of areas where emphasis was needed with time to provide help to those students with special needs. We strongly supported the development of alternative paths to algebra for students who were not ready for algebra. Although students who were not adequately prepared for a first course in algebra did not do well in that course, we learned, somewhat to our surprise, that many students significantly improved their prealgebra skills by staying in the course. We therefore recommended to those schools with no alternative routes to algebra that they encourage their students, including those whose test results indicated they were not ready for algebra, to take it and, if necessary, repeat it (preferably taught in a different way).

Although a mastery level of 70% is used to advise students that they appear well prepared in topics on all four of the diagnostic tests, that level is too high to be a realistic threshold level for determining adequate preparation. We have never tried to determine such a level. When teachers insist on using test performance as a criterion for placement in a course, we encourage them to use it only with other criteria including teacher recommendation. We also encourage them to complete an analysis in their own schools of test results compared with class performance.

Because the tests are designed to provide diagnostic information to students and teachers and because the tests have been validated by comparison with future class performance, the tests are criterion referenced rather than norm referenced. Computing and publishing norms would encourage users to compare themselves with those norms rather than to interpret the results in terms of the standards that are set by the design of the tests. These are the two major reasons why we have resisted repeated requests for computing and publishing any kinds of statewide averages. Many teachers and district mathematics specialists understand these reasons and respect our decision.

To assist departments and districts in considering their own curriculum and course plans, we will provide summary data to schools and to districts. These data include item analyses and topic and test averages for the entire school or the entire district. Reviews of these results are often instructive in identifying both programmatic successes and specific areas where new approaches or more emphasis is needed. We are eager to work with districts in interpreting these reports. We will not provide disaggregated data to schools or districts or other agencies. This helps to prevent another kind of abuse of the tests.

Program evaluation based upon comparison of test results would be an abuse of the diagnostic tests for several reasons. The tests were not designed for that purpose. They do not cover all, or even most of the topics of a course, so they are not comprehensive. The tests measure adequate preparation through a demonstration of competence with processes and concepts; they do not assess mastery of material. Our refusal to compute or distribute averages helps to prevent this possible abuse. It seems likely that our success in resisting this misuse of the tests has been a significant factor in teacher acceptance of the tests. The state-mandated tests of the California Assessment Program (CAP), which are specifically designed to compare programs and schools, have relieved the pressure for the diagnostic tests to be used for those purposes. The CAP tests do not provide results for individual students, but reporting is done in a way that encourages comparisons among schools and districts.

## IMPACT OF THE MDTP PROGRAM

The diagnostic testing program has had many effects on mathematics educators and mathematics education in California. Its primary service to teachers and students comes directly from the nature of the program's tests and the reports. Diagnostic tests provide a fairly clear way to identify each student's degree of preparation for further study and to indicate particular areas where additional preparation will be most helpful. Careful summary reports provide helpful indications to teachers of their students' levels of preparation in specific areas and of the extent to which their students have developed skills and understood concepts. Individual student reports not only help guide the students, but they are also effective in communicating some of the students' successes and needs to parents and counselors. Since the tests and the reports carry the imprimatur of the state's universities, they often help support a teacher by providing a seemingly more authoritative documentation of that teacher's judgment.

One of the most important effects of the diagnostic testing project has been to bring high school and university faculty together to address their shared concerns about the need for improving the mathematical sophistication of all students. The involvement of university faculty demonstrated to high school faculty their commitment to dealing with these issues. Similarly, the willingness of high school teachers to use the tests, to analyze the results provided, to encourage colleagues to do the same, and to attend conferences and confront curricular and pedagogical issues revealed their enthusiasm and dedication to bringing mathematics to all students. Meeting together provides exchanges of views and perspectives that are invaluable in understanding the realities of both the high school and university mathematical experiences of our students. Each should influence the other. We are trying to use the algebra readiness test as a focus for communication between senior and junior high school mathematics. As soon as it was released, high school faculty started encouraging its use in their feeder junior high schools.

User conferences provide excellent opportunities for project personnel to learn about teacher and school responses to the tests. Many teachers maintain attention on required skills and topics throughout the year by using more integration of topics and persistent reviews. One teacher, with district support, developed a large set of worksheets based upon the topics on the tests. Students are offered worksheets on those topics where their test results indicate a need for more work in order to be adequately prepared for the next course. The project has served as a clearinghouse in providing information about the worksheets to other schools and teachers, who may purchase them at a nominal cost from the sponsoring district.

Another district developed a special summer program to help its students to prepare for their next course. Its summer school calendar lists the test topic to be reviewed each day. Students are encouraged to attend only those sessions dealing with topics in which review was indicated on their MDTP test result letters. Care is taken to ensure that the same topics are covered at neighboring schools during different weeks so that students can attend at times convenient to their family vacations. Since there are no grades, the program is readily viewed by all as a means to help the students.

In some high schools, the calculus readiness test is used to obtain an indication of the preparedness of their students who plan to start a calculus course. After testing the students near the end of a precalculus course, teachers encourage some students to complete a focused review during the summer and advise others to take a more appropriate course the following year. Dramatic improvements in some advanced placement calculus courses and in the success rates of students in those courses have been reported to us from some schools.

At least one school district sends MDTP test results to parents with an accompanying letter from the district superintendent explaining the nature of the test and its significance. The letter includes a tear-off to be returned to school to ensure that the student delivery service works. It also includes an invitation to call the district office with comments or concerns. Almost universally, the response has been extremely positive. Parents are pleased to see specific recommendations of ways to help strengthen their children's mathematics education, even if the parents are not familiar with the topics mentioned. We have been told that the most common response is "how can I help."

An independent evaluator did a comparative study of four pairs of schools that were matched for socioeconomic status and racial composition of their students. One school in each pair made consistent use of MDTP materials while the other did not. A study was made of changes in mathematics enrollment and performance on the twelfth-grade CAP mathematics tests over a two year period. In each pair of schools there was a greater increase (by a factor between two and five) in the percentage of students enrolling in three or more years of mathematics courses at the MDTP school than its matched nonuser school. In all but one of the pairs, the MDTP school showed a greater increase in the percentage of students enrolling in advanced mathematics courses. The twelfth-grade CAP scores increased in each of the MDTP schools although they decreased in three of the four nonuser schools. There was a greater increase in the score of the MDTP school than of its paired nonuser school that had an increase. Of course, the MDTP schools probably have more united and committed faculties than the paired non-MDTP schools, but the tests are a resource for those faculties and can help bring faculties together.

Another effect of the MDTP program has been to make some university mathematicians more keenly aware of some of the forces shaping mathematics education and some of the issues facing it. By working with educators at other levels, by examining test results, and by listening to the concerns of committed teachers, we have become more sensitive to the needs of both teachers and students. Our involvement with the broader education community has led several of us to work directly on educational, curricular, pedagogical, and policy issues that affect all of our students. Student learning is greatly influenced by their attitudes and by their social or cultural settings. We must be aware of and responsive to these influences. Teachers need our support to maintain their commitment to mathematics. University mathematicians can help by working with teachers in institute settings similar to those of the California Mathematics Project, by working with policymakers at the state and district level to create frameworks and other statements that guide the mathematics curricula in the state, by service on accreditation and review teams, and by working with colleagues involved in teacher preparation activities. In all these settings mathematicians can strengthen the education of all students by helping to ensure that account is taken for both the demands of the discipline and the needs of the students.

## WHAT DO THE TESTS TEST?

Our understanding of what we are testing has broadened. It is easy to look quickly at the items from a mathematician's perspective and conclude that diagnostic tests simply measure manipulative skills. Indeed, some critics maintain that this is an inherent limitation of multiple choice mathematics tests. MDTP tests could not work as well as they do if they were only tests of procedures because understanding is also a critical prerequisite for continued success in the study of mathematics. The workgroup is beginning to explore ways to report test results from this perspective to supplement its existing reports along a content-oriented axis.

The following item taken from the algebra readiness test reveals how performance on items may reveal procedural or conceptual errors.

In the figure shown to the right, what fractional part of the circle is shaded?

(A) $\frac{4}{15}$ (B) $\frac{3}{8}$ (C) $\frac{5}{8}$ (D) $\frac{11}{15}$

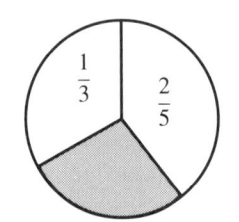

To answer this item correctly, students need to interpret the question fully, correctly add the two fractions with different denominators, and subtract that result from 1—recognizing that the entire circle is represented by the fraction 1. Students who select the last option appear to have correctly added the two fractions but to then have made one of three errors. They may have simply seen that their calculation lead to an offered response and stopped. Alternatively, they may have failed to read the question completely and have not even seen what was asked. Lastly, they may not have known that the solution required subtracting from 1. The first two errors may be test-taking errors, while the third reveals a weakness in conceptual understanding. The first two errors may also indicate students' belief that mathematics consists of exercises to be done quickly without care or thought.

Students who mark the second option compound the possible errors just listed by incorrectly adding the two displayed fractions. This error in addition of fractions, $\frac{a}{u} + \frac{c}{v} = \frac{a+c}{u+v}$, is shockingly common to college instructors but not surprising to high school teachers. Students who select the third option indicate an understanding of the underlying concepts about fractions although they also display a serious procedural weakness in addition of fractions, at least in this context. It would be interesting to see whether those students make the same kind of error when asked to simply add two fractions.

Another error that is amazingly persistent appears on the MDTP tests at every level because students' calculations continue to reveal their implicit belief that $(a+b)^2 = a^2 + b^2$. On the algebra readiness test, 17 is commonly marked as the value of $\sqrt{5^2 + 12^2}$. Similarly, many students mark $9x^2 + 4y^2$ as the value of $(3x+2y)^2$ on the test measuring readiness for intermediate algebra. On the calculus readiness test, many students choose the distractor $4x^3 + 2ax^2$ for the value of $\sqrt{16x^6 + 4a^2x^4}$.

Why do students continue to make these kinds of errors in spite of repeated instruction to the contrary? When I first saw my own students making such errors, I thought they were almost hopelessly underprepared for college work. Although this may be correct, it now seems to me that we need to understand how so many students can fail to be influenced by repeated efforts to correct such errors. To what extent do these errors result from student unwillingness to accept mathematics as an intellectual activity requiring understanding? To what extent do such errors reflect student belief that mathematics is a collection of tricks and, indeed, that the most straightforward techniques work often enough to always be used? It seems to me that we have to recognize the issues raised by these kinds of questions and respond constructively to them in order to significantly improve mathematical literacy.

## Closing Comments

We must find effective ways to help students develop the mathematical tools, knowledge, and perspectives they need. Reading and writing are essential components of learning and doing mathematics. Of course, students need to be competent and comfortable with algebraic manipulations in order to develop understanding of almost any college mathematics. They also need to have an understanding of the reasons those manipulations work and, perhaps more importantly, to have placed mathematics in a reasonable conceptual framework. They must recognize that mathematics is a subject to be understood and a means to develop reasoning skills rather than being an algorithmic activity to generate desired responses for good grades.

Diagnostic and prognostic tests, which have been carefully developed and validated, are valuable tools for improving students' mathematical education and for forging cooperative efforts of high school and university faculty to strengthen mathematics education. When properly used, the tests are an important aspect of student assessment. They must be created by knowledgeable experts in the teaching and practice of mathematics. Teamwork is essential to provide the variety of perspectives and options needed as well as to stimulate the careful and exhaustive reviewing that is essential to developing good multiple choice tests. The credibility of good diagnostic tests is established by field testing and by voluntary teacher and school adoptions. The rapid generation of reports with diagnostic data for students and teachers makes fairly detailed and objective information available to them. Certainly other forms of assessment are also important for students, teachers, and programs. However, it should be stressed that multiple choice diagnostic testing is more than an assessment instrument; it is an established resource that enhances communication among universities and secondary schools.

**Information on obtaining MDTP materials.** The tests and scoring software developed by MDTP are available for use by academic agencies outside the state of California. Review copies of the tests and a license agreement are available from the author. There is a nominal charge for permission to reproduce the tests and a copy of the software. Please contact:

>Alfred Manaster
>University of California, San Diego
>Department of Mathematics 0112
>9500 Gilman Drive
>La Jolla, CA 92093-0112

## APPENDIX

**Typical questions from each competency area of the precalculus test.**

1. Elementary operations with numerical and algebraic fractions.
$$\frac{3x-2}{x+2} - \frac{2}{x-2} =$$
(A) $\frac{3}{x+2}$ (B) $\frac{3x-4}{x^2-4}$ (C) $\frac{3x}{x^2-4}$ (D) $\frac{x(3x-10)}{x^2-4}$ (E) $\frac{3x(x-4)}{x^2-4x+4}$

2. Operations with exponents and radicals.
$$\frac{x^{3a+2}}{x^{2a-1}} =$$
(A) $x^{a+3}$ (B) $x^{a-3}$ (C) $x^{5a+1}$ (D) $x^{a+1}$ (E) $x^3$

3. Linear equations and inequalities.
For what value of $t$ does $\frac{2t-1}{3t+4} = 2$?
(A) $-6$ (B) $-\frac{9}{4}$ (C) $\frac{3}{2}$ (D) $\frac{9}{4}$
(E) There is no value of $t$ satisfying this equation.

4. Polynomials and polynomial equations.
If $(x-1)(x^2-4) + 2(x-1)(x+2) = (x-1)P$, then $P =$
(A) $x^2 - 2$ (B) $x^2$ (C) $x(x+2)$ (D) $x^2 + 2$ (E) $(x+2)^2$

# DIAGNOSTIC TESTING 37

5. Functions.

   If $f(x) = 2x + 5$ and $g(x) = 1 - x^2$, then $f(g(2)) =$

   (A) $-3$  (B) $-1$  (C) 1  (D) 2  (E) 9

6. Trigonometry.

   If $\sin \theta = \frac{3}{5}$ and $0 \leq \theta \leq \frac{\pi}{2}$, then $\tan \theta =$

   (A) $\frac{3}{2}$  (B) $\frac{4}{3}$  (C) $\frac{5}{4}$  (D) $\frac{4}{5}$  (E) $\frac{3}{4}$

7. Logarithmic and exponential functions.

   $\log_3 27 =$

   (A) 81  (B) 9  (C) 3  (D) $\frac{1}{3}$  (E) $\frac{1}{9}$

8. Word problems.

   If $\frac{2}{3}$ is $\frac{1}{2}$ of $\frac{4}{5}$ of a certain number, then that number is

   (A) $\frac{15}{4}$  (B) $\frac{5}{3}$  (C) $\frac{5}{6}$  (D) $\frac{5}{12}$  (E) $\frac{4}{15}$

DEPARTMENT OF MATHEMATICS, UNIVERSITY OF CALIFORNIA, SAN DIEGO, LA JOLLA, CALIFORNIA 92093

# Creativity: Nature or Nurture?
# A View in Retrospect

ARNOLD E. ROSS

It is quite usual for scientific or mathematical talent to manifest itself at an early age—often in the early teens. In the instances of the successful maturing of such young talent and of the development of high competence, one finds often the continued opportunity for contact with good mathematical and scientific ideas and with people who are capable of providing encouragement and guidance toward significant challenges. It appears that very vivid, early impressions leave their mark upon the nature of the ultimate achievement.

In what follows I must attempt to describe my own involvement with the very young in the last few decades and to mention the work of some colleagues of whose efforts I became deeply aware through personal association.

The involvement of the Hungarian mathematical community early in this century through its school journal, *Matematikai lapok*, and through thoughtful and friendly attention given by its members to the youngsters captured by the Journal, resulted in a dramatic flowering of mathematical and scientific talent in Hungary.[1] One needs mention but a few names, such as George Polya, Gabor Szego, John von Neumann, Theodor von Karman, Paul Erdos, Paul Turan, L. Szilard, E. Teller, and E. Wiegner, to be forcefully reminded of this fact. Most of these accomplished men ultimately made vital contributions to our own country's well-being.

Following his final return to the USSR from Cambridge in 1934, Peter Kapitza was instrumental in creating an intensive program of a national search for young talent. A. N. Kolmogorov and I. M. Gelfand contributed vitally to this effort. They were joined by many talented colleagues in the USSR.

---

[1] A. E. Ross, *Fostering scientific talent*, Science & Technology Policies, Yesterday, Today and Tomorrow, Ballinger Publishing Company, Cambridge, MA, 1973, pp. 72–77. Also see the forthcoming report by Vera John-Steiner and Reuben Hersh, *A visit to Hungarian mathematics*.

## Growing Up

My own growing up was affected by the unusual events of the revolution and civil war in the USSR and the resulting economic dislocations. As if to compensate for material privations, I was learning my mathematics in a private gymnasium organized by temporarily unemployed university professors. Thus I learned my geometry through a deep involvement in exploration and justification, a mode of upbringing not unnatural in mathematics. This form of upbringing has been referred to in later years in the United States as the Moore method.

One of the elder statesmen of Russian mathematics just before that time was S. O. Shatunovsky in Odessa, USSR. An accomplished mathematician and a charismatic teacher, he obtained special permission for four youngsters to attend university lectures. At about sixteen years of age, Feliks Gantmakher and I were the oldest of the four. We attended Shatunovsky's course in analysis in which limit theory used directed sets (the concept which is credited to him)[2] and also his course in Galois theory in which we were taught group theory simultaneously as the need arose. We attended other lectures as well. Material privation of the year of famine was made easier to bear through the excitement of our mathematical adventure.

In 1925, about a year after I returned to Chicago (my birthplace) I was able to work in Professor E. H. Moore's course in General Analysis. As we were expected to do all the proofs and even offer occasional conjectures, this was an incredibly engrossing experience. I was fortunate to be able to do this as an undergraduate; for when I graduated, Professor Moore retired. His classroom was the birthplace in the United States of the now famous Moore Method, continued with great dedication by his pupil—another Moore (R.L.).

My interest in number theory grew and by the time I graduated I could help Professor L. E. Dickson as his research assistant. I wrote my thesis in number theory under him.

## Working with the Very Young

My teachers, by their example, inspired an interest in working with the very young, which has lasted all of my life. Problems associated with this interest came in all shapes and sizes. The manifestations of these problems have rarely been trivial.

Some events one recalls more vividly than others. Establishing rapport with young people with unhappy early experience in mathematics has always been thought-provoking. Thus gaining the confidence of student nurses and apprentice technical laboratory staff at St. Louis University Medical School is a heartwarming remembrance. Turning nondescript mathematics for premedical students into an honors course (Notre Dame) was an example of

---

[2] *The mathematical intelligencer*, no. 1, vol. 11, 1989. An account is written by Allen Shields.

enlightened policy. Selecting about sixty students out of a very large group destined for remedial mathematics (The Ohio State University (OSU)), students who were afraid of mathematics and disliked it but who were extraordinarily good in everything else, and providing them with a thoughtful mathematical experience which changed their expectation of boredom and frustration into an activity long remembered with pleasure, was a victory of principle over unhappy practice. Years later in the restless early seventies a dramatic variant[3] of this last was developed for the young fellow citizens in the inner city and their children.[4] (See Editor's note for descriptions of the four programs mentioned above.)*

Watching for some manifestation of mathematics and scientific talent has been an ever present consideration. At first this led to the development of challenging honors programs for undergraduates, to the concern over the intellectual climate for the older students in the graduate school and ultimately to the development of a challenging intensive precollege summer program for the very young. Describing this last development is the principal aim of this paper.

---

[3] A. E. Ross, *The shape of tomorrows*, Amer. Math. Monthly 77 (9) (Nov. 1970).

[4] A. E. Ross, *Horizons unlimited*, preprint, 1972.

*Editor's note. In 1936, Professor Ross developed a mathematics course to replace the elementary mathematics requirement for students in the nursing and laboratory technician programs at St. Louis Medical School. Professor Ross' involvement in these programs came about from his association with a student nurse, who was about to drop out of the program because she was failing the mathematics requirement for the third time. Professor Ross was impressed by the student's competency in life sciences and he was confident that she could do mathematics. In working with this student, Professor Ross was able to help her to do mathematics thoughtfully and to become free of the fear of mathematics which had stymied her past attempts to learn mathematics.

In 1945, Professor Ross developed a quality honors course for medical students at the University of Notre Dame, after he became aware of the inadequacy of the courses in the medical school in preparing premedical students to do the kind of mathematical thinking that is required of doctors. His insights into the deficiencies of doctors' mathematical reasoning were hard won through contacts with doctors during his wife's crisis in suffering a nearly fatal toxic reaction to penicillin.

In 1964 at The Ohio State University, Professor Ross viewed the academic records of 1000–1500 students enrolled in remedial mathematics courses and discovered that sixty or so students were honor students, who excelled in all their courses except mathematics. In considering a special mathematics course for these outstanding students, Professor Ross was concerned to dispel their fear and dislike for mathematics. The course he designed, which was *not* remedial, was structured as a combination of lectures and problem seminars. He observed that the students were reserved and wary during the first weeks, but around the third week of class they would begin to show enjoyment and excitement about learning mathematics.

During 1969–1970, Professor Ross helped to develop a program for high school dropouts in their early adult years, 19–25 years of age. The purpose of the program was to educate these young adults so that they could pursue new careers. For the mathematics component of the program, Professor Ross divided the time between lectures and laboratory work. Again, the program was challenging rather than remedial. The enthusiasm of the students sparked the idea that a similar program should be available to their children. The result was a "Sunday School" for school children. The success of the program as perceived by the community it served can be seen in the name that was eventually chosen for the program, Horizons Unlimited.

When I came to Notre Dame in 1945, the intellectual climate at Notre Dame reflected the vitality of mathematical and scientific life throughout our land in the postwar years. Notre Dame was transforming itself from a famous sports-oriented college into a major university. Charles Misner, who was a physics major, became the first mathematics honors student. His early deep involvement in mathematics was reflected in his career as a theoretical physicist and cosmologist.[5] He is a co-author of the fundamental tract, *Gravitation*, with John Wheeler and Kip Thorne.

Shortly afterwards came Tom Banchoff. He was already interested in geometry. This interest influenced his work in mathematics, both pure and applied. He is well known for his highly popular computer generated movies of phenomena associated with 4-dimensional geometry. Tom thoughtfully assisted with some nontrivial, instructional experimental programs.

Other gifted young people were involved over the years. I have always felt that working in individually-designed, challenging honors programs and having an opportunity to share the responsibility for exploratory programs in science and mathematics education should be an important part of the growing up of every scientist.

To place one's preoccupation with the search and development of mathematical and scientific talent into proper social perspective, one must recall the lethargic atmosphere prevailing in our secondary schools before the Second World War. Admiral Nimitz took us to task, in his famous speech[6], for our neglect of our responsibilities—neglect which became so apparent vis-à-vis the urgent need of technological mobilization. Our educational enterprise failed us. Our engineers were all too often not equal to the technical challenges that were facing us, and the urgently needed tasks had to be performed by mathematicians and physicists. These were in short supply, and if not for the influx of talent from abroad, we might have faced insurmountable difficulties.

The tragic events in Europe, before the Second World War and their lingering aftereffects from 1935 onward, brought to our shores an incredible constellation of talent and academic wisdom. I was reminded of this when I discovered a report entitled, *To teach how to teach how to do*, written in 1962. It was exciting to see that the list of faculty colleagues in our programs included B. A. Amira (Jerusalem ), R. P. Bambah ( India), S. Chowla, Max Dehn, and Abraham Goetz (Poland), H. D. Kloosterman, Kurt Mahler, W. W. Rogosinski ,Thoralf Skolem, Father Ivo Thomas (Black Friars, Oxford), and Hans Zassenhaus. The list of talented colleagues who helped in some way is much longer.

With the war being over, we soon forgot our trials and tribulations until another rude reminder—this time from the USSR in the form of Sputnik.

---

[5] *Origins*, The Lives and Worlds of Modern Cosmologists, Alan Lightman and Roberta Brewer, Harvard University Press, Cambridge, MA, 1990.

[6] The Association of American Universities, *Journal of proceedings and addresses of the Forty-Third Annual Conference*, The University of Chicago Press, Oct. 1941, p. 92.

In order to enhance the quality of our competitiveness, great material resources were thrown at the problem of improvement of our science education. Many of our able university colleagues became deeply involved in precollege programs of various quality of ambition, varying from remedial efforts to energetic attempts to seek out and develop young talent. Not many of these programs have survived the era of complacency which followed the period of frantic post-Sputnik activity.

Following the appearance of Sputnik, great support (much greater than currently available!) was given to enhancement programs for teachers. Some of these were limited to summers and some were in place during the whole academic year. At Notre Dame we had both the summer and the academic year programs significantly interacting with each other. Many of their graduates acquired sufficient momentum to go on to higher degrees. Almost all of them took part in a great variety of experimental programs.

Through a happy accident, a small group of youngsters benefitted from the ongoing teacher program. In the spring of 1957, a number of parents phoned me expressing a concern that their able and active fifteen- and sixteen-year-old children had nothing worthwhile to do in the coming summer. I fitted these youngsters into the ground level of our (multilevel) teachers' program. The young newcomers performed exceedingly well.[7] The fact that the ideas forming the heart of the curriculum designed for the teachers were accessible to our very young audience changed, profoundly, the attitude of the teachers. It helped to overcome teachers' skepticism about the appropriateness of the curriculum designed for them. This effect could be observed very vividly in the summers of 1988 and 1989 when we revived our teachers' program.[8] (See Appendix B.)

A transition from a traditional academic interest in the able youngsters, to a very engrossing and demanding personal involvement, occurred in almost imperceptible stages. The way led from discussions with youngsters' teachers and parents to counseling the students, to working with about half-a-dozen gifted boys in the summer of 1957, and to establishing in the academic year 1957–1958 a Sunday class, dubbed "Sunday School" by the youngsters, and open to all able children within the one-hundred-mile radius of Notre Dame. Both the Sunday School and the summer term programs were continually growing in size and scope. Of the two, the summer program survived and moved with me to OSU (1964), visited the University of Chicago at the invitation of Felix Browder for four summers (1975–1978), and returned to OSU where it is today. Among our University of Chicago colleagues who shared in our work at Chicago were Professor Paul Sally, Professor Irving Kaplansky, Professor Robert Gerosh, and Professor Jonathan Alperin.

---

[7] A. E. Ross, *Notre Dame's 1960 summer program for gifted high school children*, The Mathematics Teacher (Oct. 1961).

[8] A. E. Ross, *The teacher as a role model*, preprint, 1990.

## Interaction with Colleagues Abroad

Over the years I have had an opportunity to share our experience with colleagues abroad.

I participated in the India (pilot) Program of three weeks duration (All India Summer Institute in Mathematics, Bangalore, 1973) directed by Professor V. Krishnamurthy of Barla Institute of Technology and Science in Pilani. We were able to bring within the reach of our students an impressive selection of topics in elementary number theory thanks to the assistance of our colleagues, Professor Krishnamurthy himself and Professor T. Sonndarajan of Madurai University.

Some selected topics in combinatorics were discussed by Professor Krishnamurthy. Professor K. Vankatachahengar gave a number of talks on the work of Ramanujan, and Professor M. Vankataraman discussed the relationship of mathematics with the sciences.

Lessons of the pilot program were reported at the United States-India Binational Mathematics Congress which met in Bangalore that summer, with a strong recommendation that a National Mathematics Program for the gifted should be established in India.

A strong appeal to the government was revived by the Ramanujan Mathematical Society which conceived of such a program as a fitting memorial to Ramanujan.[9]

The Australian Summer (January) Program in Canberra is of two weeks duration. It has been directed since 1969 by Professor Larry Blakers of the University of Western Australia and, I am glad to say, is still very much alive. I was a part of this program from 1975–1983. We were using number theory as a systematic introduction to mathematical thinking. Problem-solving has provided deep student involvement.

The program in Heidelberg was directed by Professor Peter Roquette in the summers of 1978 and 1979. It was of four weeks duration and was in session from mid-August to mid-September. Cooperation and warm encouragement from our colleagues at the University were very important for the success of the program. Number theory served as an environment for exploration, and the challenge of problem-solving induced deep student involvement. In its second summer, the program had two levels. I participated in the program both summers.

## Objectives

A great deal of soul searching took place in 1960 $\pm\varepsilon$ as we were trying to decide upon the scope of the youngsters' summer studies. We found that

---

[9] A. E. Ross, *On the logistics of a talent*, Proceedings of the Ramanujan Centennial International Conference, December 15–18, 1987, Annamalainagar, pp. 161–167.

whenever the selection of participants[10] uses some measure of their potential, in as much as it can be judged through the quality of their initiative, intensity of their preoccupation, and the degree of their perseverance, one is bound to net a group representing a wide spectrum of interest and temperament. At the outset, therefore, we were confronted with the dilemma as to what purpose should be served by a program for a collection of young individuals who have in common only eagerness, curiosity, an unbounded (and hitherto undirected) supply of vitality, and, possibly, an ultimate destiny in science.

Even at that time, mathematics and science had begun to permeate the work of most of the professions and were becoming a vital part of our increasingly sophisticated technological society. It became important to respond to this change in environment. Nevertheless, in doing this we had to take into account that our charges were too young to be sure of the choice of a career. Therefore, we had to make an effort to open for them as many doors into the future as possible and guard against creating an undue influence which would narrow their outlook.

We have settled on the objective of providing a vivid apprenticeship to a life of exploration. We were governed by the knowledge that science floats on a sea of mathematics. Therefore, the basis of our program has been intensely mathematical. We have felt that the education of future explorers should encourage the kind of involvement that develops the capacity to observe keenly, to ask astute questions, and to recognize significant problems. This is important for two reasons. First, the progress of every science depends upon the capacity of its practitioners to ask penetrating questions and to identify important problems. Second, we believe that personal discovery is a vital part of the learning process for every individual eager to gain deep insight into his subject.

## THE RANGE OF IDEAS

Since 1957, we have used number theory as the basic vehicle for the development of the students' capacity for observation, invention, the use of language, and all those traits of character which constitute intellectual discipline. We have chosen number theory because of its wealth of accessible, yet fundamental and deep, mathematical ideas and for its fund of challenging but tractable problems. In our treatment of number theory, we make use of the fact that it has been the birthplace of modern abstract algebra and provides the underpinning of the study of combinatorics and of discrete mathematics, generally, as well as an introduction to many ideas of modern abstract mathematics and to the theoretical tools of sophisticated, contemporary technology.

Naturally, number theory presents one with an embarrassment of riches. However, in selecting (out of a great multitude of possibilities) ideas and

---

[10]Cf. descriptive application materials for more detail. They are available upon request.

methods for our studies in number theory, we have chosen those ideas that have had the greatest impact in mathematics, both pure and applied.

Problem-solving serves as the means of achieving deep student involvement. We take advantage of the fact that problems may serve not only as exercises in the use of acquired techniques and in the development of heuristic skills, but they can and should also be used as a device which helps to achieve deeper understanding of new ideas. Our use of problems in this last manner was noted by Arthur Engel.[11]

## MATERIAL SUPPORT

In 1958, shortly after the National Science Foundation (NSF) introduced its teacher enhancement programs, it began to support precollege summer science training programs (SSTP). Its staff at that time consisted of able and dedicated middle career college and university colleagues. They believed that it would be important to encourage innovation and experimentation in the programs which they supported, and they found a way of doing this in spite of the official "guidelines" which were not very perceptive and often counterproductive, but always present in official government transactions. They watched very carefully the progress of programs for which they were responsible by arranging evaluation visits by talented and accomplished academic colleagues, who were recruited in our country as well as abroad, to serve as examiners. We benefitted from these visits and remember them with much pleasure.

The NSF provided us with some financial support in the summer of 1958 and with our first program grant in the summer of 1959. The NSF support continued through the summer of 1974 in spite of the reported criticism of some of the new staff that nurturing creative people is "elitism" in the all too popular even though tragically wrong usage of this word. The program was supported by the NSF when the program moved to the University of Chicago for four summers (1975–1978). We did not get NSF support when we returned to Ohio in 1979. Support for the summer programs was discontinued altogether in 1980. In the summer of 1988, the NSF introduced summer programs across the nation under the title "Young Scholars." Its objectives seemed to differ from those of the earlier SSTP programs. We at OSU were denied support by the NSF in 1988, but we received partial support for the summers of 1989 and 1990.

## PROGRAM STRUCTURE

Our program has been multilevel. This makes it possible for some of our participants to come for another summer and build upon the mastery of

---

[11] A. Engel, *Mathematische Schulerwettbewerbe*, Jahrbuch Uberblicke Mathematik, 1979. Wissenschaftsverlag Bibliographisches Institut, Mannheim.

ideas acquired in the preceding summer. This can be done for a number of successive summers. Each time the students would benefit from opportunities designed especially for our program participants or by attending interesting graduate courses for which they would be ready. Such advanced participants are always welcomed and encouraged by our colleagues. An accomplished and sympathetic faculty is very important for the success of a program such as ours. Such a faculty does provide an imaginative academic leadership.

No less important is the contagious enthusiasm of the nucleus of very able returnees and the dedication of the young "elder statesmen" who serve as counselors. We are deeply indebted to these young people without whom we could not achieve the high measure of participant involvement so essential to our success. Our counselors are recruited from among our experienced and able program alumni studying in the best universities in our country.

We have observed the deep and lasting impression which the sharing of the responsibility for the success of our program has made upon our young counselor-colleagues. We feel that each summer our success is due in great measure to their dedication, to their technical competence, and to their deep sense of responsibility for their charges. Their task is technically demanding, for we are able to do much nontrivial mathematics with our gifted young charges. It also demands a constant association, practically around the clock in the same living quarters, with ambitious and bright youngsters less experienced than themselves. The kind of apprenticeship, which our young scholar-mentors undergo, does, we feel, encourage the development of the best qualities of academic citizenship.

A youngster of proven talents and dedication brings to the success of a program at least as much as he carries away with him. Thus the best interests of all the participants, and not only the best interests of the most accomplished among them, are well served by our practice of returning a small group of able participants for a second, third, or even a fourth summer. We have found that these returnees contribute immeasurably to the esprit de corps of the whole group by their own example, by the quality of the competition they provide for the ablest newcomers, and by the fact that they serve as an example of the kind of recognition which we accord to outstanding achievement.

Since in the selection of participants we look for talent and creativity rather than experience, we have had to face the problem of the diversity in the background of the students we get. This problem is not resolved by restricting participants to a given age level or to a designated secondary school grade. Indeed, such artificial restrictions diminish the effectiveness of the program in a very significant way. Rather, an effective response to the unavoidable variations in experience, temperament, or thrust of major interest, which is usual in a group of talented youngsters, calls for the application of a great deal of effort and ingenuity in carrying through the academic program in the classroom as well as outside the classroom. This effort must be made, and its effect makes it well worth the pains that must be taken. Over the years the age range broadened to include youngsters from thirteen through sixteen years old and has represented all four years of high school.

The diversity of students' backgrounds has encouraged communication through sharing of experience. Their pride in achievement and their strong faith in the value of independent involvement has controlled the temptation to limit oneself to imitation. Teamwork is not discouraged. Avid conversation begins on the first evening of their arrival following the orientation session on Sunday night, which precedes the first lecture on the following morning. All of the program newcomers are given a reasonably accessible Problem Set #0 (see Appendix A), which is due the next day. Some fundamental terms and some notation are new to most of them. The only way to decipher the questions is through mathematical conversation between our newcomers and the program veterans.

## THE FIRST SUMMER'S EXPERIENCE

Many colleagues have asked for our collection of our number theory problem sets with the view of using them in their ongoing experimental programs. To use these sets effectively, one must observe that the majority of these problems are a part of the *Leitmotiv* with several strands which weave through the content, surfacing every so often as a part of some fundamental denouement brought forth by a buildup to which students themselves have contributed.

Let me illustrate this. At first we assume manipulative familiarity with integers. Description of $Z$ tightens as students' experience grows. This includes awareness of the intuitively acceptable well-ordering principle and its direct use until they understand its relation to induction. Experimenting with linear Diophantine equations and divisibility properties, they arrive at the unique factorization theorem in $Z$. To get quickly to interesting questions, congruences are introduced informally through divisibility. Calculating multiplicative inverses modulo $m$ leads to the recognition of algorithms based on the experience, acquired in the meantime, with continued fractions.

Experience with binary relations involving the transition from a graph (Bourbaki) to a description in words and vice versa, and an appreciation of the quotient space concept comes on about the tenth day of our working together.

To bring out the reasons why we speak of mathematics as "abstract" we use problems describing the transition from the "algebra of classes" to the Boolean algebra.[12] In the case of quadratic reciprocity we preface the final denouement by a discussion of the phenomenon of condensation of language leading from common parlance, to special phrases (sports!), to technical terminology (the sciences and the professions), to mathematical symbolism.

---

[12] A. E. Ross, *Towards the abstract*, Mathematical Spectrum **10** (3) (1977–1978).

Awareness of technique of generalization leads our very young explorers to establish the structure of rings of polynomials over the prime field $Z$ modulo $p$, then to study residue class rings modulo $m(x)$ and from that to asking good questions about finite fields, in general, and resolving these questions on their own.

Experience with commutative groups, occurring naturally in number theory, leads to questions about their structure. Of particular interest is the fact that the multiplicative group of a finite field is cyclic. This is studied in a slightly more general context of a finite subgroup of the multiplicative group of any field.

Intuitive feeling for real numbers makes it possible to study approximation of irrationals by rational numbers using continued fractions and to raise the question of "the best approximation." Experience with continued fractions makes it possible to explore their role in the study of Pell's equation.

Gaussian integers serve as an opening into the study of other quadratic arithmetics, involving arithmetics which are Euclidian, those which are not but have unique factorization, and some in which unique factorization fails. We make use of the well-behaved arithmetics for the solution of certain Diophantine equations. Although we come close to ideal theory, we have not done it formally. I feel that if we should do this, then we must move in the direction of extracting pertinent information from the structure of the ideal class group. If this could be made accessible to our young audience, it would give a more accurate feeling for the status of our arithmetic art. This fundamental issue competes, however, for attention with other desirable experience.

The formal definition of the derivative of a polynomial holds true for any coefficient field and allows one to view the Taylor theorem as an immediate generalization of the binomial theorem and, thence, leads to the expected properties of the derivative and to an understanding of the relationship of these to the behavior of roots. In particular, one obtains algorithms for expansions into partial fractions that are more natural than those used in the calculus instruction.

Combinatorial and geometrical arguments often allow one to avoid lengthy computations, and they lead more quickly to the heart of the matter. Determinant as the area of a parallelogram (volume of a parallelotop), Pick-Steinhouse formula for the area of a polygon with points of the fundamental lattice as vertices, and other lattice point arguments lead to quick insights into many arithmetic phenomena.

Some accessible topics in geometry of numbers, including applications to the behavior of the sum of two and four squares, some facts about the arithmetics of quaternions, and $p$-adic methods with some discussion of the solution of congruences modulo $p^n$ as an analogue of approximation to real numbers usually use up the allotted time.

One feels that flexibility is desirable so that more emphasis may be given in one summer to a topic which may be treated more lightly in the summer following. Also, a desirable new topic may be introduced unexpectedly in an exploratory fashion.

Our grown-up program veterans of many years standing will be aware of the difference in the course structure as reflected in the titles of the first summer courses. However, even though in the first summer the main burden of activity is concentrated now in one course (Number Theory), the nature of student involvement is, if anything, deeper and broader in scope than before.

Students do between twelve and twenty problems per day. Subheadings in the daily problem sets emphasize our efforts to point out the cultural content of the experience of exploration. Let me mention a few of these: "To think deeply of simple things," "Prove or disprove and salvage if possible," (never 'prove' alone!), "Numerical Problems—Some Food for Thought," "Technique of Generalization," "Exploration," etc. Examples of our problem sets and examinations are provided in Appendix A.

Each summer we have our basic course for newcomers which we call Number Theory. It begins each morning with a lecture-discussion. New problem sets are distributed at the end of each lecture. Problem seminars for small groups of participants, directed by faculty members, meet three times per week usually in the early afternoon. Studying and working on problems are done at the dormitory. Each counselor assumes the responsibility for a small group of newcomers, corrects their problem sets, and encourages discussion. Advanced participants contribute much to this last activity. A sense of vital community in which the youngsters take pride develops rather quickly. The spirit thus engendered is remembered by them vividly over many years.[13]

## ADVANCED PART : BEYOND THE FIRST SUMMER

Those of our program participants who return for another summer are given a strong course in combinatorics that is rich in problems. The first combinatorics course in our program was taught by Professor Thoralf Skolem in 1962 (Notre Dame). At OSU our colleagues in combinatorics have provided such a course for us since 1964. In the last five summers this course was taught by Professor Dijen Ray-Chaudhuri.

Other advanced courses usually vary from summer to summer. To be acceptable they must be accessible to our advanced participants and rich in content as well. Our choice depends upon the interests of the available accomplished colleagues sympathetic to the program.

Let me illustrate what we have been able to do with a few examples: In the summer of 1980 we put together a triple-track experimental course in analysis consisting of (1) a course in Real Calculus (Track I, MWF), taught

---

[13] N. D. Fisher, *The Ross Young Scholars Program at OSU*, MER Newsletter, vol. 2, no. 1, Fall 1989.

by Professor Bogdan Baishanski; (2) a course in $P$-adic Calculus (Track II, MWF), taught by Professor Kurt Mahler of the Institute of Advanced Studies Australian National University, Canberra, Australia; and (3) a problem seminar, directed by Professor Ranko Bojanic (TThr), where problems from both Track I and Track II were discussed and where the students could observe the similarities and the essential differences between these two important versions of calculus. Track II dealt with much new elementary material from the second edition of Professor Mahler's Cambridge Tract.

In addition, a course in Algorithmic Mathematics (MWF) was taught by Professor Hans Zassenhaus of OSU. Dr. David Y. Yun (a program alumnus) of the IBM Watson Research Laboratory (Chairman of the Department of Computer Science at Southern Methodist University since 1984); Professor Charles Saltzer of OSU; and Dr. M. Pohst of Köln (West Germany) participated in this course, each discussing topics pertaining to his own special interests in the subject.

In 1983, a course in Projective Planes was taught by Professor Daniel Hughes of Queen Mary's College, London University, England. Also the following two mini-courses, each of about four weeks duration, were designed for the advanced program participants: (A) Some Fundamental Ideas of Topology by Professor Dan Burghelea (OSU) formed a sequence of lectures introducing the students to the spirit and methods of "TOPOLOGY." The Euler-Poincaré characteristic of polyhedra, an important numerical measure of topological complexity, formed the central theme; (B) Motions in Hyperbolic Plane Geometry and the Classification of Surfaces by Professor Guido Mislin (ETH Zurich) served as an introduction to non-Euclidean geometry, the Poincaré model of the hyperbolic plane, Mobius transformation groups of hyperbolic translations, and classification of compact surfaces.

Two graduate courses were opened to our very advanced participants. A course in Fourier Series was taught by Professor Ranko Bojanic with special emphasis on the estimates of the rate of convergence and other approximation theoretical properties. Also a course in Analytical Number Theory was taught by Professor Barry Cipra.

In 1986, an Introduction to Representation Theory of Finite Groups designed for our advanced participants was taught by Professor Dan Burghelea. Our chemistry colleague, Professor Bruce Bursten, described how a chemist makes use of the group character theory in the study of the behavior of substances with a known molecular structure.

Also, a course in Probability was taught by Professor Antoine Brunel (University of Paris) for our advanced participants. He began by discussing discrete probability models and their combinatorial properties, then considered continuous probability models by assuming the existence of densities and using intuitive analogies with volumes to replace the technicalities of calculus. This made possible the discussion of such sophisticated applications as random walks, central limit theorems, Brownian motion, and fractal sets.

In 1987, Professor Bogdan Baishanski gave an Introduction to the Mathematics of Quantum Theory which was designed to prepare a group of our advanced participants for the three lectures given during the last week of the program by Professor Charles Fefferman based on his paper, *N-body problem in quantum mechanics.* The topics for the course were chosen in consultation with Professor Fefferman. Professor Robert Mills gave six lectures, the purpose of which was to have a physicist introduce the students to the conceptual content of quantum theory and to the relationship between mathematical structure and the actual character of quantum phenomena in physics. As a part of the introductory discussion, Professor Gerald Edgar gave two lectures on Statistical Mechanics. The first lecture dealt with the philosophy of the subject; the second lecture discussed the Ising model in one, two, and three dimensions.

Professor Ray-Chaudhuri's course in Algebraic Coding Theory and Professor Robert Stanton's Introduction to Compact Operators on Hilbert space were open to our experienced advanced participants.

In 1988, an Introduction to the Theory of Convex Sets was taught by Professor Helmut Rohrl of the University of California at La Jolla. Its content included elementary properties of convex sets, Brunn's Theorem, support lines, support hyperplanes, ovals of constant breadth, Barbier's Theorem, Radon's Theorem, Helly's Theorem and Helly-Type Theorem, applications of Helly-Type theorems, Caratheodory's Theorem, Blaschke's Selection Theorem and applications, extremum problems, isoperimetric problems, symmetrization, and geometric inequalities.

Elementary Fractal Geometry was the topic of a course by Professor Gerald Edgar. Starting with a background in calculus, the advanced participants were encouraged to write proofs (and in some cases formulate the statements) related to fractal geometry as introduced by Mandelbrot. Topics included background in metric topology and measure theory, topological dimension, Hausdorff dimension, self-similar sets, and discrete-time dynamical systems.[14] The course served as preparation for a visit by Dr. Mandelbrot who gave three lectures near the end of the summer.

Occasional lectures by vivid expositors provided much stimulation. Let me mention a few colleagues who contributed in this manner over the years: A. A. Albert, Alfred Brauer, Hans Jonas, Walter Ledermann, Wilhem Magnus, E. J. McShane, Marston Morse, Paul Rosenbloom, W. Warwick Sawyer, John Todd, and Herbert Vaughan.

Since 1983 we created a regular series of visiting lectures dedicated to Bertha Halley Ross. In 1984 Professor Felix Browder gave a lecture entitled, "Mathematics and Science." In 1986 Professor Victor Klee lectured on "Some Unsolved Problems in Discrete Mathematics." Professor Charles

---

[14]Gerald A. Edgars, *Measure, topology, and fractal geometry*, Springer-Verlag Undergraduate Texts, Springer-Verlag, Berlin and New York, 1990.

Fefferman gave three lectures on "Mathematics of Quantum Theory" in 1987. In 1988, Professor Benoit Mandelbrot gave the following three lectures on Fractals: Lecture 1, "The Mathematics and the Physics and Also the Art;" Lecture 2, "Some Simple Mathematical Topics;" Lecture 3, "Some Applications, Including Price Variations." In 1989, Professor De Witt Sumners from the Institute of Mathematics and Molecular Biology at Berkeley gave the following two lectures: "Using Knot Theory to Analyze DNA Experiments" and "Random Knots and Polymer Entanglement". In 1990, Professor Charles Misner gave three lectures forming an overview of cosmology: Lecture 1, "A Map of the Universe," and Lectures 2 and 3, "Some Problems in Mathematical Cosmology."

## Mathematics and Science

Mathematics and science have never moved far apart for long in their development. Many scientists and mathematicians, Von Neumann among them, argued for the importance of their interaction. It is thought provoking that pragmatic practitioners of either have been compelled by events to move from one to the other.

The work in high energy physics abounds in examples of this last phenomenon. The use of the character tables of finite groups by stereochemists is comparatively recent. Discovery by French solid state physicists of *P*-adic numbers in the work of Hensel is very recent. (This last discovery was reported vividly by John Maddox in *Nature*.) Knot theory, as well as differential geometry, provides increasingly fruitful insights in molecular biology. Combinatorics and discrete mathematics in general supply many more examples of interaction. The list of such events is long and growing.

We have said that we develop the capacity for scientific thinking, although remaining at the outset entirely within mathematics, by challenging the student to observe, conjecture, put his or her conjectures to the critical test of possible counterexamples, and finally to develop a convincing justification for the surviving conjectures. Such a progression of involvement represents a way of life for every scientist, not only for a mathematician. Mathematics, if approached properly, provides an early accessible opportunity for the development of the capacity for scientific thinking in the very young.

Our advanced participants are more mature than most youngsters of their age group. Because of their early start, they will be doing serious work much earlier than most of their peers. Whatever their ultimate commitments prove to be, it is highly desirable for them to get at least a glimpse of what the contemporary scientific life is like. Normally this may not happen to them until graduate school. This is too late. Therefore, we feel that each summer we should attempt to exhibit some significant contemporary interaction between mathematics and science. What we do in this endeavor, as well as the details of the design of our overall summer activities, has evolved over a period of many years.

Our collaborative course describing the use of group character tables in stereochemistry is an example of our efforts to bring out, vividly, the interaction between mathematics and science. We first tried this during our program's visit to the University of Chicago in 1975 with warm encouragement and help from our distinguished chemistry colleague there, Professor Gerhart Closs. Slightly different in structure were our ambitious efforts with quantum mechanics in 1987 and with fractal geometry in 1988. The imaginative introduction to probability by Professor Brunel (1986) provided much excitement as well as valuable experience. Very thought provoking was Algorithmic Mathematics (1980) taught by Professor Zassenhaus. It illustrated many current concerns in mathematics vis-à-vis computers. We hope to build upon our experience with putting the real calculus side by side with the calculus in ultrametric spaces (1980) and to be able to make understandable the newly discovered interactions of this last with physics.

In 1989, we were able to illustrate growing interaction between knot theory and molecular biology through an Introduction to Knot Theory by Professor Helmut Rohrl and an illustration by Professor De Witt Sumners of the use of these ideas by molecular biologists.

The kind of work which we have been doing in our program has had to be done in the face of very meager available material resources. Our program survived in spite of the financial difficulties because of the warm support by our department and by virtue of the help and warm encouragement of our accomplished colleagues. The NSF's support in the formative years of the program's existence was very important. Also important have been gifts from the friends of our program and the small occasional scholarship grants from industry obtained through intervention of colleagues in related departments.

## EVALUATION IN DEPTH

Everyone admits that working with able and eager youngsters is extraordinarily satisfying. However, many people ask if the very great effort called for and the required expenditure of public treasure are justified by the "value added" to the lives of the young program participants and through that to the welfare of the society at large.

The approbation given to the program by the Mathematical Assosiation of America (MAA) (1986) and by the American Association for the Advancement of Science (AAAS) (1988) has been heartwarming. Nevertheless, many colleagues have felt that an extensive and reasonably detailed evaluation of the effect of the program's thirty-two years of existence would be desirable. We are proceeding with such a study in collaboration with Dr. Ben Ami Blau, recently retired from IBM, and Dr. James Murphy (biostatistics) of the University of Colorado Medical School. The problem of finding the present whereabouts of about 1500 of our program alumni is nontrivial, and we are grateful for any help in this which we can get.

On the basis of incomplete responses to date, we may provide the customary not too subtle statistics: The fifty-four undergraduates, who wrote to us, study at Harvard, MIT, Yale, Princeton, Caltech, UCLA, Stanford, and UC Berkeley. Their studies include mathematics, physics, chemistry, biophysics, biochemistry, computer science, engineering, molecular biology and economics.

The twenty-four graduate students, from whom we have heard, study at Harvard, MIT, Yale, Princeton, Caltech, UCLA, Stanford, UC Berkeley, and the University of Texas. They study mathematics, physics, theoretical physics, chemistry, biophysics, biomathematics, molecular biology, computer science, and economics.

Of the eighty-five alumni with a Ph.D. whom we have reached, seventy-three are in academic life (research and/or teaching) and twelve have responsible positions in industry. They represent a wide spread over the years going back to 1957. Their professional interests include mathematics (both pure and applied), logic, physics, chemistry, astrophysics, geophysics, computer science, operations research, medicine, and English.

In addition to the above, we have reached a number of accomplished non-Ph.D. alumni holding responsible professional positions in industry or business enterprises.

At the moment there are hopeful indications that the quality of professional activity of our program alumni is high. Among our alumni polled to date, is included a recipient of the Gold Medal of the Faraday Society (chemistry), a recipient of the distinguished Rossi Award in High Energy Astronomy, three Young Scientist Presidential Awards, more than fourteen recipients of Sloan Fellowships, at least one recipient of a MacArthur Fellowship, and a significant number (twenty-four) of other recipients of national and international postdoctoral research fellowships and grants.

The seventy-eight alumni of our younger generation who responded to the survey have received a combined total of eighty-three fellowships and grants from their universities, the NSF, and other foundations.

## Prospect

Creativity provides vitality to our society. In the past when the pace of change was more deliberate, we could afford to let manifestation of this vital facet of life spring up through the random juxtaposition of happy influences. We cannot live with this comfortable state of affairs in our competitive and rapidly changing technical society. We should recognize the need for an imaginative intervention which could bring forth the desperately needed outcome.

As a society, we have signaled loud (if not clear!) our satisfaction with our possession of the cadres of singular achievement—our Nobel Laureates—and our concern over the prevalent illiteracy in most matters, including scientific. However, we have failed to appreciate and be concerned over the need for large intermediate cadres of talented, well-trained, resourceful individuals capable of imaginative initiative.

Happily there are harbingers of change. We are beginning to be aware of the fact that developed talent is a natural resource of vital importance to the national well-being. Also, we have come to recognize that genuine search for talent must be carried out with the broadest possible social and economic base. This last strategy is dictated by the fact that the springs of talent are widely dispersed and are no respecters of geographic, social, or economic boundaries.

Much remains to be done. Opportunities for broadly-based collaboration are growing. I hope that I will still be able to continue to share in this growing, heartwarming enterprise.

# Appendix A:
# To Think Deeply of Simple Things

SSTP Number Theory.   Problem Set #0    OSU Ross  6/17

**Reading Search:**
Question 1. What are matrices? How does one add and multiply them?
Q2. What is a perfect number?
Q3. What is a Mersenne prime?
Q4. What is the meaning of $\sum_{d>0}^{d|N} d$ ?

**Asking Good Questions:**

P1. Compare the following mathematical systems with each other:
$$\mathbb{Z}, \mathbb{Q}, \mathbb{R}, 2\mathbb{Z}, \mathbb{Z}_3, \mathbb{Z}_6, \mathbb{Z}_8, \mathbb{Z}_{11}.$$
There are two operations (addition and multiplication) in each of these systems. Which of these systems resemble each other in regard to the essential properties of these operations?

**Exploration:**

P2. Consider a two by two matrix $\alpha = \begin{bmatrix} 0 & 1 \\ 2 & 0 \end{bmatrix}$ whose elements are in $\mathbb{Z}_3$. Next consider the mathematical system
$$\mathbb{Z}_3[\alpha] = \{aI, bI+b\alpha, c_0 I + c_1\alpha + c_2\alpha^2, \ldots \}$$
where $I = \begin{bmatrix} 1 & 0 \\ 0 & 1 \end{bmatrix}$ and the $a$'s, the $b$'s and the $c$'s are in $\mathbb{Z}_3$. Thus our system consists of all the polynomials in $\alpha$ with coefficients in $\mathbb{Z}_3$. How many distinct elements are there in our system? How does it compare with $\mathbb{Z}$? With $\mathbb{Q}$? With $\mathbb{Z}_{11}$? With $\mathbb{Z}_7$? With $\mathbb{R}$?

**Numerical Problems** (Some Food for Thought).

P3. Let $N = 10723$. Write $N$ to base ten, to base three, to base two, to base 3, to base eleven.

## PLATE 1. Problem Set #0.

**P4.** In the last case introduce, if necessary, the next digit $t = \text{ten}$, $e = \text{eleven}$. Without change of base (a) add $(673)_7$ and $(3545)_7$, (b) subtract $(2436)_7$ from $(4534)_7$, (c) multiply $(562)_7$ by $(345)_7$, (d) divide $(5062)_7$ by $(34)_7$. Here the base is 7 throughout.

**P5.** Using division to base 7, write $N = (34652)_7$ to base 2, to base 5, to base eleven.

**P6.** Write each of the following numbers to base 3: 3, 9, 27, 243, $\frac{1}{3}$, $\frac{1}{81}$. Write each of these numbers to base 2.

**P7.** Find the following elements in $Z_5$: $-1, \frac{1}{2}, \frac{2}{3}, \sqrt{-1}$. How many of these elements can you find in $Z_6$? In $Z_{10}$? In $Z_{11}$? In $Z_{13}$?

**P8.** Calculate each of the following summations: $\sum_{d|16} \frac{1}{d}$, $\sum_{d|128} \frac{1}{d}$.
$$\sum_{d|496} \frac{1}{d}.$$

Can you make an interesting conjecture? Perhaps one more example may help:
$$\sum_{d|496} \frac{1}{d}.$$

Calculate this last sum and compare its value with the values of the first two sums.

# OSU Number Theory Problem Set #2  A.E. Ross  Columbus  6/24

**Reading Search:**

**Q1.** If $\ell \in \mathbb{R}$, what is "the greatest integer function $[\ell]$"? Calculate $[\sqrt{57}]$, $[-\sqrt{57}]$, $\left[\frac{\sqrt{7}}{7}\right]$, $[2+\sqrt{8}]$.

**Q2.** What is meant by the canonical factorization of a positive integer into prime factors?

**Exploration:**

**P1.** Consider complex numbers $a+bi$ with $a \in \mathbb{Z}$ and $b$ in $\mathbb{Z}$. These numbers (Gaussian integers) form under addition and multiplication a mathematical system (denoted by $\mathbb{Z}[i]$) not unlike $\mathbb{Z}$. Which of the properties in P9–P10 of $\S$SNT hold true also in $\mathbb{Z}[i]$? Prove or disprove and salvage if possible.

**P2.** $n \in \mathbb{Z}$ and $2 \nmid n \Rightarrow 8 \mid n^2-1$.

**P3.** $a \mid b$ and $a \mid c \Rightarrow a \mid br + cs$ for every $r$ and $s$. True in $\mathbb{Z}$. True in $\mathbb{Z}[i]$.

**P4.** Let $a$ and $n$ be positive integers greater than $1$. If $a^n - 1$ is a prime then $a=2$ and $n$ is a prime.

**Numerical Problems** (Some Food for Thought).

**P5.** Construct a table of "logarithms" (indices) for $U_{17}$. Use the table of "logarithms" in P5 to find all the solutions of each of the following equations in $\mathbb{Z}_{17}$: a) $x^2 = 2$; b) $7x^2 = 6$; c) $x^3 = 3$.

**P7.** Find all the generators in $U_{19}$. Find all the non-zero elements in $\mathbb{Z}_{19}$ which are perfect squares in $\mathbb{Z}_{19}$. Here $U_{19}$ denotes the group of units in $\mathbb{Z}_{19}$.

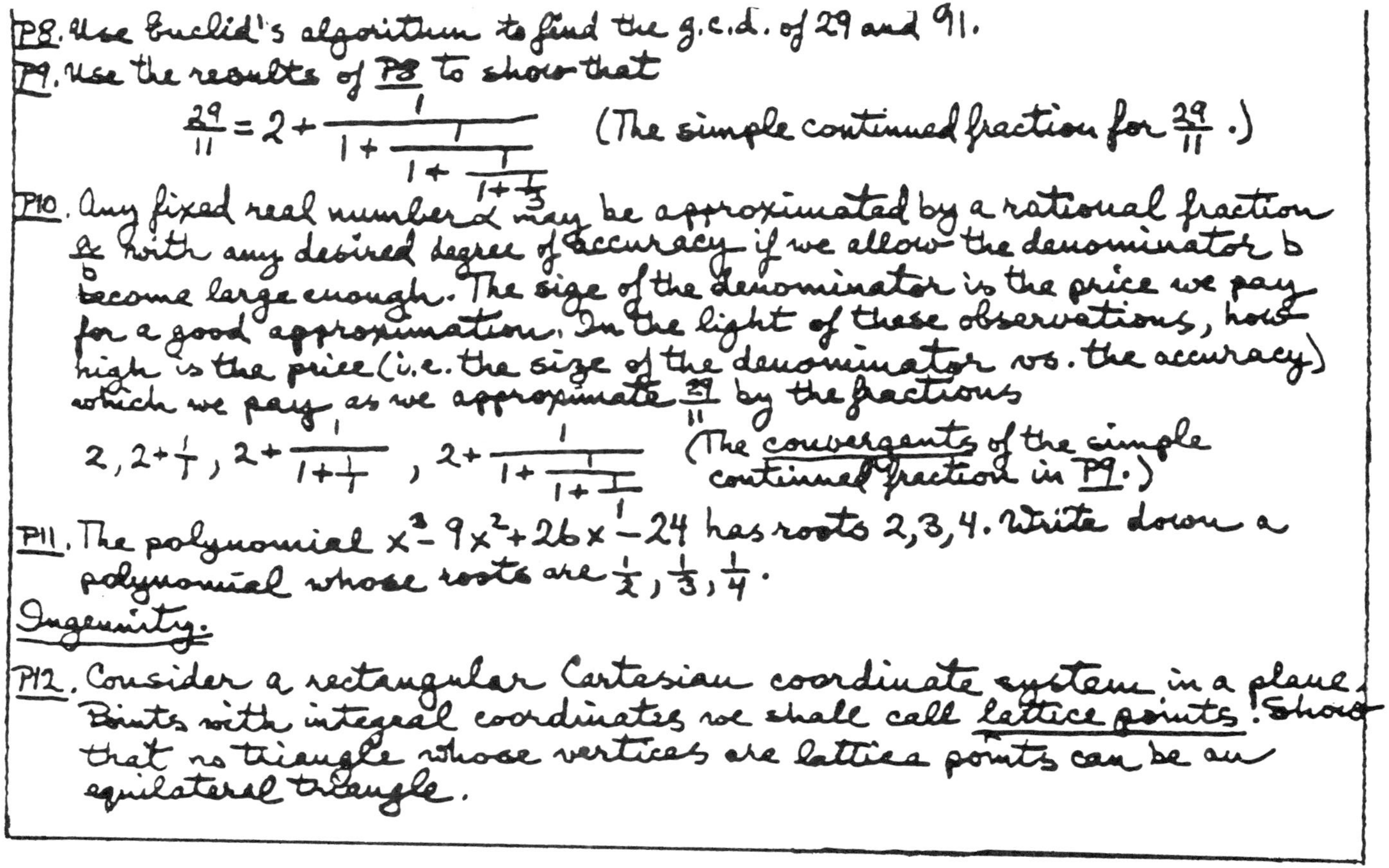

P8. Use Euclid's algorithm to find the g.c.d. of 29 and 91.

P9. Use the results of P8 to show that
$$\frac{29}{11} = 2 + \cfrac{1}{1+\cfrac{1}{1+\cfrac{1}{1+\cfrac{1}{3}}}}$$
(The simple continued fraction for $\frac{29}{11}$.)

P10. Any fixed real number may be approximated by a rational fraction with any desired degree of accuracy. If we allow the denominator to become large enough. The size of the denominator is the price we pay for a good approximation. In the light of these observations, how high is the price (i.e. the size of the denominator vs. the accuracy) which we pay as we approximate $\frac{29}{11}$ by the fractions

$$2, \; 2+\frac{1}{1}, \; 2+\cfrac{1}{1+\frac{1}{1}}, \; 2+\cfrac{1}{1+\cfrac{1}{1+\frac{1}{1}}}$$

(The convergents of the simple continued fraction depicted in P9.)

P11. The polynomial $x^3 - 9x^2 + 26x - 24$ has roots $2, 3, 4$. Write down a polynomial whose roots are $\frac{1}{2}, \frac{1}{3}, \frac{1}{4}$.

Ingenuity:

P12. Consider a rectangular Cartesian coordinate system in a plane. Points with integral coordinates we shall call lattice points. Show that no triangle whose vertices are lattice points can be an equilateral triangle.

PLATE 2. Problem Set #2.

OSU Number Theory Problem Set #4  A.E.Ross Columbus 6/26

### Reading Search:
Q1. What is the highest power of a prime contained in a factorial?

### Exploration.
P1. We have considered a rational function of the form

$$\cfrac{1}{a_1 + \cfrac{1}{a_2 + \cfrac{1}{a_3 + \cdots + \cfrac{1}{a_n}}}}$$

of the $n$ variables $a_1, a_2, a_3, \ldots, a_n$

and we have called this function a finite continued fraction. For this continued fraction we use the convenient notation

$$[a_1, a_2, a_3, \ldots, a_n].$$

Set $P_1 = a_1$, $P_2 = a_2 a_1 + 1$, $\ldots$, $P_\ell = a_\ell P_{\ell-1} + P_{\ell-2}$, $\ldots$
    $Q_1 = 1$,  $Q_2 = a_2$, $\ldots$, $Q_\ell = a_\ell Q_{\ell-1} + Q_{\ell-2}$, $\ldots$

Then $\quad [a_1, a_2, \ldots, a_\ell] = \dfrac{P_\ell}{Q_\ell} \quad$ (The $\ell$-th convergent!)

for $\ell = 1, 2, \ldots, n$.

Hint: $[a_1, a_2, \ldots, a_{\ell-2}, a_{\ell-1}, a_\ell] = [a_1, a_2, \ldots, a_{\ell-2}, a_{\ell-1} + \dfrac{1}{a_\ell}]$.

P2. Using the relations in P1, show that

(1) $P_\ell Q_{\ell-1} - Q_{\ell-1} P_\ell = (-1)^{\ell+1}$ for $\ell > 1$ and (2) $P_\ell Q_{\ell-2} - Q_{\ell-2} P_\ell = (-1)^\ell a_\ell$ for $\ell > 2$.

Prove or disprove and salvage if possible.

**P3.** $a \in \mathbb{Z}$, $b \in \mathbb{Z}$ and $(a,b)=1 \Rightarrow ax+by=1$ has a solution in integers $x, y$.

**P5** $a \mid bc$, $(a,b)=1 \Rightarrow a \mid c$. True in $\mathbb{Z}$.

**A Reminder.**

**P6.** A convenient description of $\mathbb{Z}$!

**Numerical Problems** (Some Food for Thought).

**P7.** Use Euclid's algorithm to calculate $(7469, 2464)$.

**P8.** Find an integral solution $\{x,y\}$ of the Diophantine equation

$$7469x + 2464y = 210.$$

**P9.** Find an integral solution $\{x,y\}$ of the Diophantine equation

$$2689x + 4001y = 17.$$

**P10.** What is the order of 28 in $U_{29}$? Of 16 in $U_{29}$? Of $28 \cdot 16$ in $U_{29}$? Next consider $U_{71}$. What is the order of 7, of 2, of $7 \cdot 2 = 14$, of 54, of 51, of $54 \cdot 51$? Any conjectures?
To facilitate calculations one can use the tables of "logarithms" (Indices!).

**Technique of generalization.**

**P11.** Consider the set of all polynomials in $x$ with coefficients in $\mathbb{Z}_3$. Denote such a set of polynomials by $\mathbb{Z}_3[x]$. Is $\mathbb{Z}_3[x]$ a ring under the addition and the multiplication of polynomials? Is $\mathbb{Z}_3[x]$ a commutative ring? With (multiplicative) identity? With the cancellation law? When should we say that one polynomial divides another in $\mathbb{Z}_3[x]$? What are the (multiplicative!) units in $\mathbb{Z}_3[x]$? Which of the following polynomials are primes in $\mathbb{Z}_3[x]$: $x^2+2x+2$, $x^4+x^2+1$, $x^3+x^2+x+2$, $x^3-x^2+1$, $x^4+1$, $x^4+x+2$?

PLATE 3. Problem Set #4.

OSU Number Theory Problem Set #5  A.E. Ross Columbus 6/28

Prove or disprove and salvage if possible.

**P1.** $d$ is the smallest positive value of $ax+by$ for integral values of $x$ and $y \Rightarrow d = (a,b)$. True in $\mathbb{Z}$.

**P2.** $p$ a prime and $p|ab \Rightarrow p|a$ or $p|b$. True in $\mathbb{Z}$.

**P3.** If $m$ and $n$ are positive rational integers and if $m|n$, then $m \leq n$.

**P4.** If $\{x_0, y_0\}$ is an integral solution of the equation $ax+by = c$, where $a,b,c$ are in $\mathbb{Z}$, then all the integral solutions $\{x,y\}$ of this equation are given by the formulas $x = x_0 + bt$, $y = y_0 - at$ for $t \in \mathbb{Z}$.

**P5.** The parallelogram in Fig.1 is determined by the vector $\alpha$ with components $a_1, a_2$ and by the vector $b$ with components $b_1, b_2$. The area of this parallelogram is equal to the absolute value of the determinant

$$\left| \begin{matrix} a_1 & b_1 \\ a_2 & b_2 \end{matrix} \right| = a_1 b_2 - a_2 b_1.$$

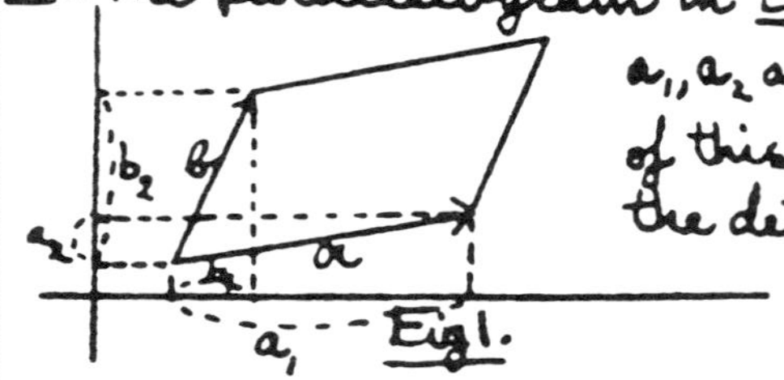

Fig.1

**P6.** Every rational integer $>1$ is a product of positive primes.

**P7.** $m|a-b \Rightarrow (a,m) = (b,m)$.

Numerical Problems (Some Food for Thought).

**P8.** Find an integral solution $\{x,y\}$ of the Diophantine equation
$$5391x + 3976y = 11$$
with the least positive value of $x$ and also one with the least positive value $y$.

P9. Expand $\frac{5391}{3476}$ into a simple continued fraction and make use of the process given in P1, Set #3 (and also in P1, Set #4) for calculating the values of successive convergents. Compare the actual value of the difference between each convergent $\frac{P_n}{Q_n}$ and the given fraction, with $\frac{1}{Q_n Q_{n+1}}$ and with $\frac{1}{Q_n^2}$. Any conjectures?

P10. Find all the positive integral solutions $\{x,y\}$ of the Diophantine equation
$$158x + 57y = 20000.$$

P11. Find all the solutions of
$$2017x = 532$$
in $Z_{4001}$. Explain.

P12. Expand $\sqrt{3}$ into a simple continued fraction using the results of P1, Set #3. How much do you know about the continued fraction for $\sqrt{3}$ after calculating, say, the first five partial quotients? Any conjectures?

P13. Multiply $2x^5 + 3x^3 + x + 4$ by $3x^4 + 2x + 2$ in $Z_5[x]$. Multiply the same two polynomials in $Z_6[x]$. In each case compare the degrees of the factors with the degree of the product. Explain.

Technique of generalization

P14. Factor those polynomials in P11, Set #4 which are not primes, into prime factors. Construct a table of primes of degree 2 in $Z_3[x]$. Does some form of division algorithm hold true in $Z_3[x]$? In $Z_{17}[x]$? In $Z_p[x]$?

P15. Problems 1 and 2 in the "Towards the Abstract" paper.

PLATE 4. Problem Set #5.

# OSU Number Theory — TEST #1 — A. E. Ross Columbus 7/3

**Reasoning:** Please indicate which properties in our inventory of basic properties of $\mathbb{Z}$ you use in each argument. Do phrase both precisely and concisely.

P1. If $m > 0$ then $m \geq 1$. True in $\mathbb{Z}$.

P2. $a, b \in \mathbb{Z} \Rightarrow$ there exist $q, r$ in $\mathbb{Z}$ such that $a = bq + r$ and $0 \leq r < |b|$.

P3. $0 \cdot a = 0$ for every $a$ in $\mathbb{Z}$.

P4. $(-a)b = \overset{?}{=} (-b) = -(ab)$. True in $\mathbb{Z}$.

Prove or disprove and salvage if possible. {Here proofs may be based on important results derived from any a reduced inventory of properties of $\mathbb{Z}$.}

P5. $a | bc$, $(a,b) = 1 \Rightarrow a | c$. True in $\mathbb{Z}$.

P6. $a \in \mathbb{Z}_m$, then $a$ is a unit in $\mathbb{Z}_m$ if and only if $(a, m) = 1$.

P7. If $a$ and $b$ are positive integers and $a | b$, then $a \leq b$.

P8. If $N(\alpha)$ is a prime in $\mathbb{Z}$, then $\alpha$ is a prime in $\mathbb{Z}[i]$.

P9. $a \in \mathbb{Z}$, $b \in \mathbb{Z}$ and $(a, b) = 1 \Rightarrow ax + by = 1$ has a solution with $x, y \in \mathbb{Z}$.

P10. $(a^n - 1, a^m - 1) = a^{(n,m)} - 1$. Here $a, n, m$ are positive integers.

P11. $a, b \in \mathbb{Z}[i] \Rightarrow$ there exist $q, r \in \mathbb{Z}[i]$ such that $a = bq + r$ and $N(r) \leq \tfrac{1}{2} N(b)$.

P12. $\alpha = a_1 + a_2 i \in \mathbb{Z}[i] \Rightarrow |(\mathbb{Z}[i])_\alpha + \alpha \cdot i| = N(\alpha)$.

P13. If $u_1 \in U_m$ has the order $w_1, w_2$, and $u_2 \in U_m$ has the order $w_2$, then the order of $u_1 u_2$ is $w_1 w_2$.

P14. A polynomial of degree $n$ with coefficients in $\mathbb{Z}_m$ cannot have more than $n$ distinct roots in $\mathbb{Z}_m$.

**Numerical Problems** (Some Food for Thought) {Please display your calculation neatly. Calculate efficiently.}

P15. Obtain the required results without changing to base ten:
(a) $(534)_6 \cdot (243)_6 = (?)_6$, (b) $(3432)_9 - (716)_9 = (?)_9$, $(4321 2)_6 \div (34)_6 = (?)_6$, $(635 1)_7 \overset{?}{=} (?)_5$

P16. Calculate the following elements in $\mathbb{Z}_{23}$: $-1, -7, \tfrac{1}{5}, \tfrac{7}{3}, \sqrt{2}, \sqrt{-7}$. Explain.

P17. Construct a table of "Logarithms" (indices) for $\mathbb{Z}_{19}$ and use it to solve the following equations in $\mathbb{Z}_{19}$: (a) $x^3 = 7$; (b) $7x = 11$.

P18. Find all the positive $(x>0, y>0)$ integral solutions $\{x,y\}$ of the Diophantine equation $223x + 89y = 713$.

P19. Expand $\sqrt{11}$ into a simple continued fraction and find the convergent with the smallest denominator which approximates $\sqrt{11}$ within $\frac{1}{1000}$.

P20. Calculate $\mu(315)$, $\tau(315)$, $\sigma(315)$. Also $\zeta(315)$

P21. Find the g.c.d. $(f,g)$ of $f(x) = x^4 - 3x^3 + 2x^2 + 4x - 1$ and $g(x) = x^2 - 2x + 3$ in $\mathbb{Z}_3[x]$.

P22. Is $1+2i$ a unit in $(\mathbb{Z}[i])_{3+5i}$? Justify your assertion.

P23. Construct the (finite) ring $(\mathbb{Z}_3[x])_{x^2-2}$. Is this ring a field? Is $x$ a generator of the group of units in this ring? Justify your conclusions. Prove or disprove and salvage if possible.

P24. For every real $\alpha > 0$ there exist infinitely many rational fractions $\frac{p}{q}$ for which
$$\left|\alpha - \frac{p}{q}\right| < \frac{1}{q^2}.$$

P25. $n \in \mathbb{Z}$, $n > 0 \Rightarrow n$ is a prime $\iff$ only if $\sigma(n) + \varphi(n) = n\tau(n)$

Problem Set (July 4). Please do all those problems on Test #1 which you did not get to during the test period. Also make whatever changes you wish to make in those test problems which you did complete.

PLATE 5. Test #1.

# OSU Number Theory Problem Set #17    A.E. Ross Columbus 7/16

Prove or disprove and salvage if possible.

**P1.** Define the symbol $\left(\frac{a}{p}\right)$ by
$$\left(\frac{a}{p}\right) \equiv a^{\frac{p-1}{2}} \pmod{p} \text{ and } \left(\frac{a}{p}\right) = +1 \text{ or } -1 \quad (\text{Legendre!}).$$

Here $p$ is a positive odd prime in $\mathbb{Z}$ and $(a,p) = 1$. If also $(b,p) = 1$, we have

(1) $a \equiv \square \pmod{p} \iff \left(\frac{a}{p}\right) = 1$;

(2) $a \equiv b \pmod{p} \implies \left(\frac{a}{p}\right) = \left(\frac{b}{p}\right)$;

(3) $\left(\frac{ab}{p}\right) = \left(\frac{a}{p}\right)\left(\frac{b}{p}\right)$;

(4) $\left(\frac{a^2}{p}\right) = 1$;

(5) $\left(\frac{-1}{p}\right) = (-1)^{\frac{p-1}{2}}$.

**P6.** $a, b \in \mathbb{Z}$ are positive, odd, relatively prime $\implies \sum_{0<x<\frac{b}{2}\atop x\in\mathbb{Z}}\left[\frac{ax}{b}\right] + \sum_{0<y<\frac{a}{2}\atop y\in\mathbb{Z}}\left[\frac{by}{a}\right] = \frac{a-1}{2}\cdot\frac{b-1}{2}$.

Hint: Draw a picture.

**P7.** $\frac{a}{b}, \frac{a''}{b''}, \frac{a'}{b'}$ are three successive entries in a Farey sequence of order $n$, then
$$\frac{a''}{b''} = \frac{a+a'}{b+b'}.$$

**P8.** $d \in \mathbb{Z}, d > 0, d \ne \square$. Then $|P_n^2 - dQ_n^2| < 2\sqrt{d} + 1$ for every convergent $\frac{P_n}{Q_n}$ to $\sqrt{d}$.

**P9.** Use the arithmetic in $\mathbb{Z}[i]$ to show that if $p$ is a positive prime of the form $4n+1$ then the Diophantine equation $x^2 + y^2 = p$ has an integral solution $\{x,y\}$.

## Numerical Problems (Some Food for Thought).

**P7.** Draw the picture used in the geometric proof of
$$\sum_{0<x<\frac{b}{2}}\left[\frac{ax}{b}\right]+\sum_{0<y<\frac{a}{2}}\left[\frac{by}{a}\right]=\frac{a-1}{2}\cdot\frac{b-1}{2}, \quad x,y\in\mathbb{Z},\ (a,b)=1$$
for the case $a=7$, $b=5$. Also for the case $a=17$, $b=11$.    (P6, Set #17)

**P8.** The congruence $x^2 \equiv 10 \pmod{13}$ has exactly two distinct solutions. Why? How many distinct solutions does the congruence $x^2 \equiv 10 \pmod{169}$ have? How many distinct solutions does the congruence $x^2 \equiv 10 \pmod{2197}$ have? Find all the solutions of each of the above congruences. Explain and generalize.

**P9.** Find all the solutions of each of the following congruences:

(1) $x^2 \equiv 2 \pmod{2737}$, (2) $x^2 \equiv 3 \pmod{1771}$.

**P10.** Does the Diophantine equation $x^2+y^2+3^3=95$ have an integral solution $\{x,y,3\}$?

**P11.** Calculate $\sum_{d \mid 3+5i} \varphi(d)$. We take $\varphi(1)=1$ and select only one d among its associates in $\mathbb{Z}[i]$.

**P12.** Construct an isomorphism between $(\mathbb{Z}[i])_7$ and $(\mathbb{Z}_7[x])_{x^2+1}$.

**P13.** Is 7 a prime in $\mathbb{Z}[\sqrt{3}]$? Is 17 a prime in $\mathbb{Z}[\sqrt{2}]$? Is 7 a prime in $\mathbb{Z}[\sqrt{5}]$? Explain.

**P14.** "Towards the Abstract" paper; Section 7, Problems 10 and 11.

Exploration.

**P15.** Find all the solutions of the congruence
$$x^4+4x^3+2x^2+2x+12 \equiv 0 \pmod{625}$$

PLATE 6. Problem Set #17.

OSU Number Theory Problem Set #20  A.E.Ross Columbus  7/23

### Exploration

**P1.** $f(x) \in \mathbb{Z}[x]$. If $f(x)$ is composite (not a prime!) in $\mathbb{Q}[x]$, then $f(x)$ is composite (not a prime) in $\mathbb{Z}[x]$.

### Numerical Problems (Some Food for Thought).

**P2.** Find a positive integer $n$ such that $\mu(n) + \mu(n+1) + \mu(n+2) = 3$.

**P3.** Is 17 a prime in $\mathbb{Z}[\sqrt{2}]$? Explain.

**P4.** Consider $\mathbb{Z}[\sqrt{2}]$. Here $34 = 2 \cdot 17 = (14 + 9\sqrt{2})(14 - 9\sqrt{2})$. Does this show that the unique factorization theorem does not hold true in $\mathbb{Z}[\sqrt{2}]$? Explain.

**P5.** $\frac{5}{7}$ is a member of Farey series (sequence) of order 11. Construct the member which succeeds $\frac{5}{7}$ in this sequence. Explain.

Prove or disprove and salvage if possible.

**P6.** If $P$ is a positive rational prime such that $P = 2^{2^n} + 1$, then 3 is a generator of $U_P$. Would this be true if $P = 2^{4^n} + 1$.

**P7.** If $n > 1$, then no two successive terms of Farey series of order $n$ have the same denominator.

**P8.** If $n$ is composite, then $\varphi(n) < n - \sqrt{n}$.

**P9.** Let $P$ be a positive odd integer in $\mathbb{Z}$ and $a \in U_P$. Extending Jacobi's generalized Legendre symbol as follows: We suppose that

$P = P_1 \cdot P_2 \cdots P_k$ (The primes $P_i$ are not necessarily distinct) and for $a$ such that $(a,P) = 1$, we set

Jacobi $\longrightarrow \left(\frac{a}{P}\right) = \left(\frac{a}{P_1}\right)\left(\frac{a}{P_2}\right)\cdots\left(\frac{a}{P_k}\right)$ ← Legendre

Then (1) $\left(\frac{a_1 a_2}{P}\right) = \left(\frac{a_1}{P}\right)\left(\frac{a_2}{P}\right)$; (2) $\left(\frac{a_1}{P}\right) = \left(\frac{a_2}{P}\right)$; $a_1 \equiv a_2 \pmod{P} \Rightarrow \left(\frac{a_1}{P}\right) = \left(\frac{a_2}{P}\right)$;

(4) $\left(\frac{-1}{P}\right) = (-1)^{\frac{P-1}{2}}$; (5) $\left(\frac{2}{P}\right) = (-1)^{\frac{P^2-1}{8}}$; (6) $\left(\frac{2}{P}\right) = (-1)^{\frac{P^2-1}{8}}$.

P10. (1) If $P, Q$ are distinct positive odd primes in $\mathbb{Z}$, then $\left(\frac{P}{Q}\right)\left(\frac{Q}{P}\right) = (-1)^{\frac{P-1}{2}\cdot\frac{Q-1}{2}}$.

(2) $P$ and $Q$ are distinct odd positive odd $(P,Q) = 1 \Rightarrow \left(\frac{P}{Q}\right)\left(\frac{Q}{P}\right) = (-1)^{\frac{P-1}{2}\cdot\frac{Q-1}{2}}$ (Jacobi).

We assume that every program participant leaves forward to do ultimately some modern science or engineering and mathematics by some other contemporary profession. Today the development of ideas moves so fast that in preparing for the future one must not look only for detailed prescription of how to solve known problems. One must acquire capacity for scientific thinking in order to meet future needs. We have tried to make this clear by our prologue, Professor Bourbaki reinforced this view in his recent lecture. Einstein expressed his view of the relation of mathematics and science in the famous quotation given by E. T. Bell in his book "Men of Mathematics".

Test #2 will take place on Tuesday, July 24 from 8:00 am to 10:00 am. 33% of the Test problems will be taken from the Sample Test 2 (Parts I and II). 33% from the Problem Sets 0-19.

PLATE 7. Problem Set #20.

OSU Number Theory    TEST #2    A.E. Ross Columbus 7/24

*Reasoning.* (Do please be both precise and concise.)

P1. $a, b \in \mathbb{Z} \Rightarrow$ there exist $q, r$ in $\mathbb{Z}$ such that $a = bq + r$ and $0 \leq r < |b|$.

P2. $0 \cdot a = 0$ for every $a$ in $\mathbb{Z}[i]$.

P3. $(-a)b = -(ab)$. True in $\mathbb{Z}[\sqrt{-2}]$.

*Prove or disprove and salvage if possible.*

P4. Let $a, b, c, d \in \mathbb{Z}[\sqrt{-2}]$ and suppose that $d$ is an "integer" of the smallest norm such that $ax + by = d$. Then $d = (a, b)$.

P5. Let $\pi \in c$ prime in $I = \{\frac{a_1}{2} + \frac{a_2}{2}\sqrt{-7} \mid a_1, a_2 \in \mathbb{Z}, a_1 \equiv a_2 \pmod{2}\}$. If $\pi \mid a^\ell$ where $a \in I$ and $\ell$ is a positive rational integer, then $\pi \mid a$.

P6. Let $a = Pq + r$, $0 \leq r < P$. Then $\left[\frac{a}{P}\right]$ is such if $0 \leq r < \frac{P}{2}$ and it is odd if $\frac{P}{2} \leq r < P$.

P7. The fundamental theorem of arithmetic holds in $I$ of P5 above.

P8. Let $H$ be a finite subgroup of the multiplicative group $F^*$ of a field $F$. Then $H$ is cyclic.

P9. If $P$ and $\gamma$ are two distinct positive odd primes in $\mathbb{Z}$, then $\left(\frac{P}{\gamma}\right) = (-1)^{\frac{P-1}{2} \cdot \frac{\gamma-1}{2}} \left(\frac{\gamma}{P}\right)$. Here $\left(\frac{P}{\gamma}\right)$ is the Legendre symbol.

P10. The congruence $g_1 x \equiv g_3 \pmod{m}$, where $g_1, g_2, m$ are in $\mathbb{Z}[i]$, has a solution $y = y_0$ in $\mathbb{Z}[i]$ iff $d = (g_1, m) \mid g_2$. The number of distinct solutions of this congruence is $(?)$.

P11. $a, b \in \mathbb{Z}[\sqrt{-2}] \Rightarrow$ there exist $q, r \in \mathbb{Z}[\sqrt{-2}]$ such that $a = bq + r$ and $N(r) < N(b)$.

P12. $u_1, u_2 \in U_m$. If $n_1$ is the order of $u_1$ and $n_2$ is the order of $u_2$ then there exists $u_3 \in U_m$ the order of which is the l.c.m. $[n_1, n_2]$ of $n_1$ and $n_2$.

P13. $\pi$ is a prime in $\mathbb{Z}[i]$, $u \in U_\pi \Rightarrow u^{N(\pi)-1} = 1$ in $U_\pi$.

P14. $\pi_1, \pi_2$ are distinct primes in $\mathbb{Z}[i]$, $u \in U_{\pi_1 \pi_2} \Rightarrow u^{(N(\pi_1)-1)(N(\pi_2)-1)} = 1$ in $U_{\pi_1 \pi_2}$.

P15. $ab \equiv ac \pmod{m} \Rightarrow b \equiv c \pmod{m}$. True in $\mathbb{Z}[\sqrt{-2}]$.

P17. $d \in \mathbb{Z}$, $d > 0$, $d \not\equiv \square$. Then $|P_n - dQ_n| < 2\sqrt{d} + 1$ for every convergent $\frac{P_n}{Q_n}$ to $\sqrt{d}$.

P18. Let $f(n) = \sum_{d \mid n} g(d)$. Then if $f(n)$ is multiplicative so is $g(n)$. True in $\mathbb{Z}$.

P19. There exist infinitely many primes in $\mathbb{Z}[\sqrt{2}]$.

P20. $\sum_{0<x\le b/2}[\frac{ax}{b}] + \sum_{0<y\le a/2}[\frac{by}{a}] = \frac{a-1}{2}\cdot\frac{b-1}{2}$, $x,y \in \mathbb{Z}$, $(a,b)=1$.

P21. $p$ is a positive prime in $\mathbb{Z}$, $\pi(x)$ a prime of degree $n$ in $\mathbb{Z}_p[x] \Rightarrow \pi(y) | y^{p^n} - y$.

P22. $\sum_{d|n, d>0} \mu(d) = 0$ or $1$.

Numerical Problems (Some Food for Thought) {Please display your calculations neatly! Calculate efficiently.}

P23. Find an integral solution $\{x,y\}$ of $x^2 - 23y^2 = 1$. P24. As in P23 for $x^2 - 23y^2 = -1$.

P25. Let $\alpha = [2,1,4]$. Find the quadratic polynomial in $\mathbb{Z}[x]$ of which $\alpha$ is a root.

P26. Find all the distinct solutions in $\mathbb{Z}[i]$ of the congruence $(3+5i)x \equiv 2 \pmod{7+11i}$.

P27. Construct the residue class ring modulo $x^3+x^2+2$ in $\mathbb{Z}_3[x]$. Is this a field? Check by calculating the number of distinct solutions without solving. What is the order of $x$ in the group of units? Explain.

P28. Is $x^2 = 1$ a unit in P27. Calculate $(x^2-1)^{-1}$. Find all the roots of $y^3 + y^2 + 2$ in $(\mathbb{Z}_3[x])\frac{3}{x^3+x^2+x}$. Explain.

P29. Find all the solutions of the congruence $x^2 \equiv 2 \pmod{119}$.

P30. As in P29 for $x^2 \equiv 3 \pmod{121}$.

P31. Find all integers $x$ such that $x \equiv 3 \pmod 8$, $x \equiv 11 \pmod{20}$, $x \equiv 1 \pmod{15}$.

P32. Calculate $\varphi(5), \varphi(7), \varphi(297)$ in $\mathbb{Z}$ and $\varphi(5), \varphi(7), \varphi(7+6i)$ in $\mathbb{Z}[i]$. Explain.

P33. The points of what lattice in the plane "represent" integral multiples of $3 + 2\sqrt{-2}$ in $\mathbb{Z}[\sqrt{-2}]$? Draw a picture and explain.

Problem Set #21. 7/24/84

Please do all those problems in Test #2 which you did not get to during the test period. Also make whatever changes you wish to make in any of solutions you did give.

PLATE 8. Test #2.

OSU Number Theory Problem Set #22  A.E.Ross Columbus  7/25

## Exploration.

**P1.** Let $f(n) = \prod_{\substack{d \mid n \\ d > 0}} g(d)$. Express the function $g$ in terms of the function $f$.

Prove or disprove and salvage if possible.

Given a polynomial
$$f(x) = a_n x^n + a_{n-1} x^{n-1} + \cdots + a_1 x + a_0 \qquad (f(x) \in F[x], \text{ Fancy field.})$$
we **define** (!) the derivative of $f(x)$ as
$$f'(x) = n a_n x^{n-1} + (n-1) a_{n-1} x^{n-2} + \cdots + a_1.$$

**P2.** If $f(x)$ and $g(x)$ are polynomials then $(f(x) + g(x))' = f'(x) + g'(x)$.

**P3.** If $a$ is a constant, $f(x)$ a polynomial then $(a f(x))' = a f'(x)$.

**P4.** Let $f_n(x) = x^n$. Then $f_n'(x) = n x^{n-1}$, $f_n''(x) = n(n-1) x^{n-2}$, ..., $f_n^{(\ell)}(x) = n(n-1) \cdots (n-\ell+1) x^{n-\ell}$.

Observe that
$$\tfrac{1}{\ell!} f_n^{(\ell)}(x) = \binom{n}{\ell} x^{n-\ell}.$$

**P5.** $f_n(x+y) = (x+y)^n = x^n + \binom{n}{1} x^{n-1} y + \binom{n}{2} x^{n-2} y^2 + \cdots + \binom{n}{n} y^n$
$$= f_n(x) + \tfrac{f_n'(x)}{1!} y + \tfrac{f_n''(x)}{2!} y^2 + \tfrac{f_n'''(x)}{3!} y^3 + \cdots + \tfrac{1}{n!} f_n^{(n)}(x) y^n = \sum_{\ell=0}^{n} \tfrac{1}{\ell!} f_n^{(\ell)}(x) y^\ell$$

**P6.** If $m > n$ we have $f_n(x+y) = \sum_{\ell=0}^{m} \tfrac{1}{\ell!} f_n^{(\ell)}(x) y^\ell$.

**P7.** Use the results in P2 and P5 to show that if $f(x)$ is a polynomial of degree $m$, then
$$f(x+y) = \sum_{\ell=0}^{m} \tfrac{1}{\ell!} f^{(\ell)}(x) y^\ell \qquad \text{(The Taylor formula.)}$$

P8. Use the result in P7 to show that
$$(f(x)g(x))' = f'(x)g(x) + f(x)g'(x).$$
What if we have the product of more than two factors? What about higher derivatives?

P9. Use the result in P8 to show that if $\alpha$ is a root of multiplicity $k$ of a polynomial $f(x)$, then $\alpha$ is a root of multiplicity $k-1$ of $f'(x)$.

P10. If $\nu_q(m) = \frac{1}{m} \sum_{\substack{d \mid m \\ d \geq 0}} \mu\left(\frac{m}{d}\right) q^d$, then $\nu_q(m) > 0$. Here $q \in \mathbb{Z}$, $q > 1$.

## Numerical Problems (Some Food for Thought).

P11. Find all the solutions of the congruence
$$x^3 + 19x^2 - x + 23 \equiv 0 \pmod{42}$$

P12. Find all the solutions of the congruence
$$x^3 - 2x + 6 \equiv 0 \pmod{125}.$$
Hint: Use the results in P7.

P13. Is 3422 a square modulo the prime 5683?

P14. For what primes $p$ is 3 a square modulo $p$?

## Reasoning

P15. There has been much discussion among members of our group about a fundamental and a very useful result in the theory of finite fields. A number of different arguments (some of which were not correct!) were put forth. The following property of polynomials has been at the foundation of every correct argument: every polynomial point is an identity in every commutative ring containing all the coefficients of the polynomials involved.

PLATE 9. Problem Set #22.

# OSU Number Theory Problem Set #27  AERoss Columbus 8/1

## Geometrical Methods.

**P1.** If $\frac{P_n}{Q_n}$ is the n-th convergent of the simple continued fraction $[a_1, a_2, \ldots, a_n, \ldots] = \alpha$,

then there exists no fraction $\frac{P}{Q}$ with $0 < Q < Q_n$ such that
$$\left|\frac{P}{Q} - \alpha\right| < \left|\frac{P_n}{Q_n} - \alpha\right|.$$

Thus $\frac{P_n}{Q_n}$ is the "best" rational approximation of $\alpha$.

**P2.** If $\left|\frac{P}{Q} - \alpha\right| < \frac{1}{2Q^2}$, then $\frac{P}{Q}$ is a convergent of the continued fraction expansion of $\alpha$.

## Numerical Problems (Some Food for Thought).

**P3.** Is 17 a prime in $\mathbb{Z}[\sqrt{2}]$? In $\mathbb{Z}[\sqrt{3}]$? Explain.

**P4.** How did you test whether the polynomial $x^6 + x + 1$ in P4 Set #24 is a prime in $\mathbb{Z}_2[x]$?

**P5.** How many primes of degree 5 are there in $\mathbb{Z}_3[x]$? In $\mathbb{Z}_7[x]$?

**P6.** Consider the convex region
$$x_1^2 + x_2^2 < 2p.$$

Find all the lattice points of the lattice in P12 Set #28 which are contained in this region. Do this for $p = 5, 7, 13, 17$.

**P7.** Consider the lattice $\{x_1 v' + x_2 v'' \mid x_1, x_2 \in \mathbb{Z}, \ v' = [2,3], \ v'' = [3,5]\}$. Does every point $(v_1, v_2)$ with integral coordinates belong to this lattice? Explain.

**P8.** **Exploration.** Consider squares $C(2m, 2n)$ of area 4 with centers at $(2m, 2n)$ as indicated in Fig 1. Draw a symmetric (relative to the origin $(0,0)$) convex figure $K$. If the area of $K$ is greater than 4, then when we translate all these squares so that $(2m, 2n)$ will go to $(0,0)$ some of the translates of $K \cap C(2m, 2n)$ must overlap!

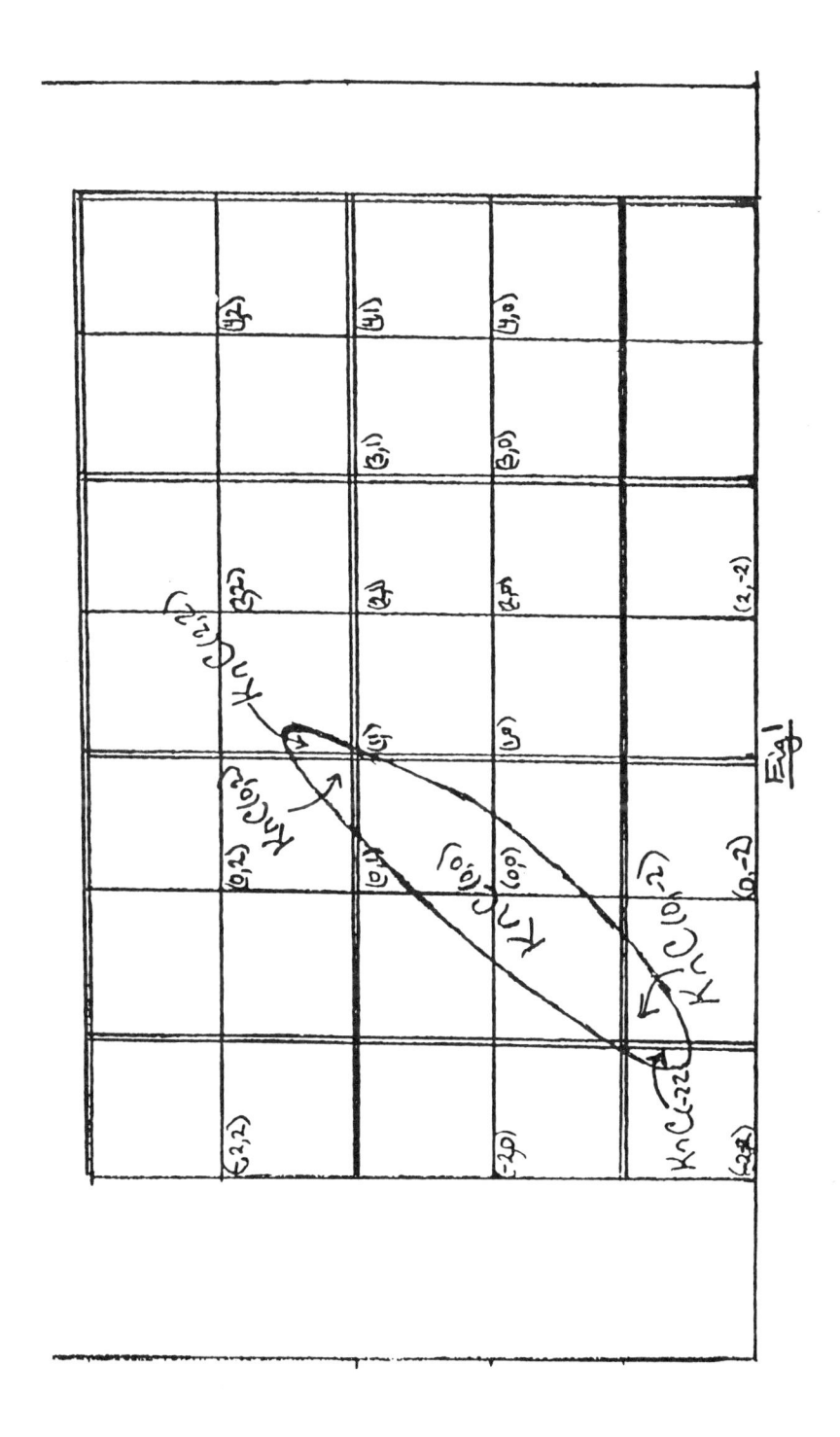

PLATE 10. Problem Set #27.

# OSU Number Theory TEST #3 A E Ross Columbus 8/8

**Prove or disprove and salvage if possible.**

P1. The congruence $x^2 + y^2 + 1 \equiv 0 \pmod{p}$ has an integral solution $\{x, y\}$ for every prime $p$.

P2. If $ab = 8^3$ and $(a,b) = 1$, then each one of $a$ and $b$ is a perfect cube. True in $\mathbb{Z}[\sqrt{-2}]$.

P3. $a, b \in \mathbb{Z}$, $a > 0$, $b > 0$, $a \nmid b$ and $b \nmid a \Rightarrow [a,b] > a$, $[a,b] > b$.

P4. If $G$ is a finite subgroup of the multiplicative group of a (commutative) field, then $G$ is cyclic.

P5. $a \in \mathbb{Z}[\sqrt{-2}]$. $N(a)$ is a prime in $\mathbb{Z} \iff a$ is a prime in $\mathbb{Z}[\sqrt{-2}]$.

P6. $p$ is a positive prime in $\mathbb{Z}$, $\pi(x)$ a prime of degree $n$ in $\mathbb{Z}_p[x] \Rightarrow \pi(y) | y - y$.

P7. What are all the units in $\mathbb{Z}[\sqrt{3}]$? Justify your assertion.

P8. If $p$ is a positive odd integer, then $x^2 \equiv 2 \pmod{p}$ has a solution $\iff (-1)^{\frac{p^2-1}{8}} = 1$.

P9. Let $x^2 - dy^2 = 1$, $d \neq \square$, $d > 8$, $x, y$ positive in $\mathbb{Z}$. Then $\frac{x}{y}$ is a convergent in the continued fraction expansion of $\sqrt{d}$.

P10. $\pi(y)$ is a prime of degree $n$ in $\mathbb{Z}_p[y]$, then $\pi(y) | y^{p^n} - y$ if and only if $n | n$.

P11. $d \neq \square$, $d > 0 \Rightarrow$ the Diophantine equation $x^2 - dy^2 = 1$ has a solution $(x,y)$ with $y \neq 0$.

P12. Any convex body $K$ symmetrical about the origin and of volume greater than $8$ contains pair of points of the fundamental lattice other than the origin. True in $\mathbb{R}^3$.

P13. The geometry of numbers to show that $p = x_1^2 + x_2^2$ has an integral solution for every prime $p \equiv 1 \pmod{4}$.

P14. If $F$ is a finite field, then $F$ is isomorphic to $(\mathbb{Z}_p[x])\pi x$ for a suitable prime $p$ and a prime polynomial $\pi(x) \in \mathbb{Z}_p[x]$.

P15. Every convergent $\frac{p}{q}$ of the simple continued fraction expansion of $\alpha \in \mathbb{R}$ is a "best approximation" to $\alpha$.

P16. If $|\frac{p}{q} - \alpha| < \frac{1}{2q^2}$, then $\frac{p}{q}$ is a convergent of the continued fraction expansion of $\alpha$.

P17. If a prime $p \equiv 1 \pmod 3$, then the congruence $x^2 + x + 1 \equiv 0 \pmod p$ has a solution.

P18. Let $\frac{p_{n-1}}{q_{n-1}}$, $\frac{p_n}{q_n}$ be two successive convergents of the expansion of $\alpha$ into a simple continued fraction. Here $\alpha \in \mathbb{R}$, $\alpha > 0$. Then $|\frac{p_n}{q_n} - \alpha| < |\frac{p_{n-1}}{q_{n-1}} - \alpha|$.

## Numerical Problems (Some Food for Thought)

**P19.** Are the fields $(Z[i])_3$ and $(Z_3[x])_{x^2+1}$ isomorphic? Justify your conclusion.

**P20.** If a prime $p \equiv 1 \pmod{3}$ then the congruence $x^2+x+1 \equiv 0 \pmod{p}$ has a solution. Justify.

**P21.** Find all the positive $(x>0, y>0)$ integral solutions $\{x,y\}$ of the Diophantine equation $261x + 109y = 849$.

**P22.** Find all the solutions of the congruence $x^3 + 3x^2 + 1 \equiv 0 \pmod{343}$.

**P23.** For what primes $p$ is $7$ a square in $U_p$?

**P24.** How many distinct solutions does the congruence $x^2 \equiv 2 \pmod{527}$ have?

**P25.** Find all the integers satisfying the following system of congruences:
$x \equiv 3 \pmod{72}$, $x \equiv 43 \pmod{108}$, $x \equiv 31 \pmod{176}$,

**P26.** The congruence $(x^2-13)(x^2-17)(x^2-221) \equiv 0 \pmod{m}$ has a solution for every $m \in \mathbb{Z}$.

**P27.** Does $f(x) = x^4 + x^3 - 3x - 5x - 2$ have multiple roots? Answer without calculating the roots. If yes, then without factoring $f(x)$ construct $g(x)$ which has the same roots as $f(x)$, with each root of $g(x)$ having the multiplicity one. Explain.

**P28.** Is there Euclid's algorithm in the arithmetic of $I = \{a_1 \frac{-1+\sqrt{-11}}{2} + a_2 \frac{1+\sqrt{-11}}{2} \mid a_1, a_2 \in \mathbb{Z}\}$

**P29.** Decompose into partial fractions in $\mathbb{R}[x]$:
$$\frac{3x+5}{x^5+4x^4+7x^3+7x^2+4x+1}$$

**P30.** Calculate the number of prime factors of $x^{292}-1$ in $\mathbb{Z}_3[x]$ of degree 1, of degree 2, of degree 3, of degree 4, of degree 5, of degree 6.

### Problem Set #30

Do please complete the discussion of those problems which you did not finish or did not get to on the test.

PLATE 11. Test #3.

## APPENDIX B: THE TEACHER AS A ROLE MODEL

In a primitive society the very young had role models close by, often in their own family. Today the teacher, increasingly, must fulfill the role of a convincing and attractive role model.

Dedicated teachers must be given an opportunity to deepen their intellectual involvement and to acquire a measure of independence which would lead to inventive improvisation in the classroom. The all too common practice of teaching the teachers procedures, which they can reproduce in the classroom verbatim, degrades the educative process. It certainly does not illustrate the vital capacity to learn through exploration.

The role model mechanism for creating a new generation of doers is often labeled *apprenticeship*. It is almost universally practiced on the level of sophisticated research. It is critically urgent, I believe, to incorporate this principle into the development of teachers, then move it into the classroom to affect our very young at the impressionable phase of their intellectual development.

We would like to present some of the experiences we have had in recent summers in reviving a teacher program developed during the Sputnik era. Our current revival turns things around from the time of the first program. It is organized as an adjunct of our summer program for able students; whereas, the summer program for the youngsters was born in the Sputnik era as an adjunct of our program for teachers.

**Clinical Practice.** The teachers are much moved by the opportunity to observe us at work with the very young in which we challenge our young charges with ideas not normally considered accessible to them.

The first teacher reaction is that of incredulity. This changes into a feeling of deep intellectual excitement and a growing conviction that they should attempt to go beyond rote practice in their own teaching. Observing us at work with students (apprenticeship!) gives the teachers courage to try to do this in their own classrooms.

One should point out that we invite participating teachers, not on the basis of the strength of their mathematical background, but on the basis of some significant indication of the high quality of their commitment to teaching. Thus the mathematical content of the program and our outlook are new to most of them.

To appreciate the teachers' experiences during the course of the program, we present selections from their own accounts. The first set of teacher commentaries describes the teachers' reactions to the program, from initial shock at our methods to acceptance. The last two commentaries describe how the teachers relate the summer experience to their own teaching.

> Immersed in mathematics! During the first few days, somehow that phrase seemed an inappropriate description of the sum-

mer mathematics institute. A more striking comparison would be to dump a non-swimmer in the deepest part of the blood-thirsty, shark-infested Atlantic Ocean and yelling to the victim, '*SWIM*'. New concepts of number theory were constantly being hurled at the struggling in hopes that an element would be used as a rescue line to stay afloat. Unknowingly the situation rapidly surfaced into a learning experience with the freedom to delve [into] and investigate the many facets of mathematics that most science teachers never question. As the experience at the institute progressed, a tremendous number of phase changes occurred as 'liquid' fundamental concepts began to 'gel' and later developed into the building blocks for a 'concrete' foundation for an even further expanded investigation.

Philip Jones
Davie, Florida


... For the first ten days of this program, I was very skeptical and convinced this program worked only on paper and in the minds of those running this summer session. The more I see of it, the more impressed I have become. The technical aspects of teaching and the pedagogy that translates into effective teaching is, many times, not easily recognized. After many weeks of watching, listening and participating, I can now see how fruitful this program and [its] methods are.

I have taught mathematics in the public high schools for 24 years. I have observed, slept, dreamt, debated, laughed, cried and loved teaching. There are many things in the teaching profession that are hard to change. Teachers are sometimes slow to change, hard to convince and slow to accept innovations in the art of teaching. They must be persuaded to accept new ideas and techniques and to try them.

[Observing the young participants] I have noticed an awakening of many of these kids. They have learned to be more investigative and eager for knowledge. These young students search for patterns, make conjectures and study long and hard. They are learning to formulate and present good questions and to discover many new mathematical concepts.

I have been convinced it works.

This has been one of the best and most intense learning experiences of my life.

Larry D. Kight
Beardstown, Illinois

As a teacher, I feel that this program has awakened me from too long a period of dormant complacency. The challenge and intensity of the program [have excited] my interest in continued education and revitalized my enthusiasm for the more difficult math courses. Learning from example can never be overstated and these last weeks have allowed me to see the best at work. Students are guided to discover non-trivial concepts naturally in a setting geared to experimentation and conjecture. The problem sets are carefully picked so that major ideas are foreshadowed early in the course. The intensity level is designed to challenge the most gifted, but still keep all working without discouragement. We are pushed to achieve at a higher level than thought possible and only in retrospect can one fully appreciate the vast strides made in understanding the subject matter.

This program is of extreme benefit for teachers because it works with a segment of the high school population that is essential for the success of this nation and that is all too often neglected, the very able student.

This summer has been the most significant educational experience that I have had for many years. If the basic premise that the future of this country is in the hands of our gifted young population and that a better teacher will help produce a better student, then programs such as this must be allowed to flourish throughout our country. I, personally, will carry the fruits of this summer's labor with me for many years.

Pat Mauch
Ekalaka, Montana


At the beginning, the experience is a frightening one as one undertakes the voyage through a mathematical maze not realizing one's own capacity to discover. But gradually, as the program proceeds, dormant abilities awaken and one's own light begins to illuminate a solid path through the work.

To accomplish such awakening, Professor Ross very strategically leads his group to the verge of conclusions through his lectures; but contrary to conventional classroom instructors, he does not provide generalizations or formulate to commit to memory. Instead, he allows his students to derive such through their own homework and mind stretching exercises.

Juan Gonzalez

After attending the 1988 Summer Math Program at OSU this teacher returned to the school with renewed spirit to implement a philosophy that had been observed as an essential component of teaching techniques. The active participation of students became an integral part of the classroom dialogue in the honors chemistry classes at Nova High School in Davie, Florida.

Through TAPPS (Talking Aloud Pairs Problem Solving), students were separated into pairs. When given problem sets of exploratory investigations, one student would become an active participant by explaining parameters and solutions to the problem. The other student would listen, giving encouragement, guidance, and criticism. The roles of the TAPPS partners reversed once a solution was developed to the satisfaction of both students.

Within a short period of time some unexpected 'happenings' occurred. First a lunch tutorial became a daily event in the classroom. During this time students openly sought peer assistance and evaluation to questions in all academic areas. But even better, students developed a spirit that chemistry was a challenge that could be met through 'sharing' experiences. Since students were given the freedom to explore aspects of the sciences beyond the confined curriculum and frameworks, a sense of 'well being' in the physical sciences developed. Also a feeling of 'community' and a science 'pep club' began to blossom during the year. With much enthusiasm, students encouraged, guided, and assisted others in various competitions. As a result, over 80% of students enrolled in advanced chemistry courses participated in events. 'Best in Shows' were won in science fairs, symposia, poster contests, essay competitions, and even a seafood cooking contest. But lastly, the overall student average on the year end final was higher than in previous years.

I am 'sold' on the active involvement of students freely investigating in the classroom. Now is the time for more teachers being flexible to let students have a greater role in the educational process. Only through the nurturing of creative and critical thinking skills will students be able to meet the 'test' of science, technology, and society at the dawn of the 21st century.


Philip Jones


When I applied to be a participant in this program I hoped to learn more mathematics. I have not been disappointed. Participation in this program has opened my eyes to mathematics as an exciting field of discovery and exploration.

What I had not anticipated, however, was the unique and efficacious aspects of the program as a vehicle for improving science teaching in general and my teaching in particular.

This program teaches science teachers how to teach science. We teach as we are taught, don't we? What is so different is that here the science of math is taught as a process. I came here regarding math as a science in the same way that philosophy is a science: a field of knowing. Instead, I found an open association of mathematicians exploring some very important objects of nature: the undefined entities and very well defined postulates and definitions which are vital to the language of the physical and social sciences.

When math is taught as a process, stress is put on working in groups. Students are stimulated to think and wonder [by way of] guided discoveries which both stimulate participants to recognize how much there is to be discovered and challenge participants to investigate further on their own. Carefully planned homework assignments stress reasoning and lead to further exploration and reinforcement of what has already been learned.

When math is taught as a process, emphasis is placed on teaching critical thinking as a result of observation, problem solving and decision making as a result of practice, and the development of insight, creative and innovative thinking. ...

When math is taught as a process ... Talent is cultivated by not putting artificially low expectations on anyone for any reason, by instructors offering encouragement and taking a friendly interest in the progress of each person, and by emphasizing the challenges rather than grading or grades.

Donald R. Sloat

# Equity and Excellence in the University of Minnesota Talented Youth Mathematics Program (UMTYMP)

HARVEY B. KEYNES

## INTRODUCTION

The University of Minnesota Talented Youth Mathematics Program (UMTYMP) is a statewide program aimed at providing an alternate educational experience for the most gifted among Minnesota's mathematically talented students. Typically identifying students in grades 6–8, the program provides an intense academic environment and a culture of mathematics via a sequence of specially-designed, accelerated mathematics courses. More details can be found in a previous volume of these *Proceedings* [1]. In that paper, we outlined the concerns of UMTYMP for improving its participation of underrepresented groups.

This paper will describe our recent approaches to dealing with the issues of equity and excellence, as well as some of the initial activities by the program to enhance female participation. We will give some of the philosophies and rationales behind the design of these activities and interventions. In addition, we will describe our plans to maintain a long-term evaluation of these interventions and measure their effectiveness. Finally, we will discuss some pilot curriculum which will provide some newer technology-based instruction and perhaps attract more students of color to participate in UMTYMP.

## THE APPROACH OF UMTYMP TO EQUITY AND EXCELLENCE

UMTYMP and Special Projects strongly adhere to the principles and connection between equity and excellence as eloquently presented in *Everybody counts* [2]. In particular, the program shaped its approach under a philosophy best expressed by the following principle [2]: "Equity for all requires excellence for all; both thrive when expectations are high." The implementation of this principle to a program such as UMTYMP with the existing high standards and levels of achievement required careful planning and design.

The statistical data from the program over its thirteen-year history and evaluations from UMTYMP graduates indicated that UMTYMP was, overall, doing an excellent job in educating its students. The postsecondary record of UMTYMP students is striking. College opportunities for both admissions and scholarships are impressive. Several major universities, including Chicago, Harvard, and MIT, give preferential admission consideration based on the prior successes of UMTYMP graduates. Female graduates fare especially well. Most choose difficult engineering and science majors at leading schools and easily persist in their programs. In alumni surveys, they testify to the importance of UMTYMP in their intellectual growth and in developing self-confidence. The program appears to patch the severe "pipeline leakage" too often seen by female students at the undergraduate level.

In the most recent survey of all graduates, 183 responded and eighty provided additional comments. Sixty-six of these comments were supportive of UMTYMP and thirty-four were highly supportive. Only six of the comments were critical. Initial data indicates that significant numbers of UMTYMP alumni are continuing with graduate studies in science, mathematics, and engineering and are considering academic careers.

With this background, the program decided that the best approach would be to maintain the historical strength of the program and curriculum and to design a variety of support, social, and counseling activities which would enhance the learning environment and atmosphere for capable students who needed these extra interventions. We wanted these students to have the same benefits and opportunities that made the program so valuable to its traditional participants. Thus, we looked at many issues concerning school and social environments, ways to improve family support, study habits, counseling needs, and, in general, to provide a more sympathetic learning environment. We were convinced that the ability to deal with mathematics at the level that UMTYMP required was present in many students who were currently not in the program, including females and students of color. Following the models of the most successful intervention programs, we wanted the new atmosphere of the program to draw out the interest and motivation of more students by having themselves discover the benefits and rewards as well as the sacrifices of the long-term culture of mathematics engendered by UMTYMP. Our first results have clearly indicated that this approach can succeed quite admirably.

## Issues Facing Female Students

Based on comparisons with other programs that involve serious selection criteria and intense personal commitment, historical participation by female students in UMTYMP has been quite good. However, considering both statistical data and anecdotal information, it was felt that greater involvement could be achieved.

Prior to 1988–1989, females comprised approximately 40% of the 1300–1500 students testing, 30% of the 100 students qualifying, and 22% of the first-year enrollments. Moreover, participation was heavily weighted toward the first two years of the program, with only about 17% overall female enrollment.

The issues associated with these statistics are quite complex, but several main features stand out. Students were primarily chosen to participate in the qualifying examination by school identification. It appears that the schools do not do as well at identifying mathematically talented female students as they do with male students. Even when a female student qualified, she was more likely to turn down admission than her male counterpart. Informal analysis indicated lack of encouragement in both the schools and the home as important factors. Once in the program, given equal ability and equal grades, female persistence was lower.

The lack of an appropriately supportive environment—in school and at home—as well as the lack of other girls in the class were important issues here. All too frequently one family indifferently requests withdrawal of a female performing near the top of her class while another tenaciously pleads to continue a boy who is struggling. Cultural issues such as socially coping with being a smart female and lack of realization of the impact of mathematics were other important aspects. Also, inconsistent and sometimes negative messages from the schools for participation in UMTYMP affected females more severely. Finally, issues affecting personal esteem—lack of self-confidence in abilities and lack of involvement in even cooperative competitions—played a prominent role.

Based on a generous and imaginative grant from The Bush Foundation, we designed a series of interventions and support activities to improve the awareness of the program and the environment for female students. We felt that we might improve female retention by creating a different level of social involvement within the program. We took the point of view that all of these interventions should be carefully evaluated and that the program should try to identify which interventions worked and for what reasons. We also took care to note attitudinal shifts and anecdotal information which also pointed to changes but which are more difficult to quantify. Because of limited resources, we planned to retain those interventions that were most effective. Although these interventions were designed to influence female students, we emphasize those that affect all UMTYMP students. We saw an important secondary effect of our intervention to be the improvement of the program for all students. Moreover, we believed that the experience developed in this type of project would enable UMTYMP to initiate, more successfully, a similar project for underrepresented groups.

## THE BUSH FOUNDATION INTERVENTION PROJECT

With initial support, starting September 1988, from The Bush Foundation, the program designed several interventions addressing primarily informational, counseling, and support issues rather than the program structure itself. They involved working with the families and the schools, as well as with the students. We believe that the need to encourage a supportive home and school environment cannot be overemphasized.

A statement on parental responsibility endorsed recently by both the National Council of Teachers of Mathematics (NCTM) and the Mathematical Association of America (MAA) provides an excellent rationale for encouraging parental and school involvement in the intervention process. Distributed to UMTYMP parents, it asserts that "social, economic, or educational status of parents does not have as important an effect on their children's learning as what parents do with their children." We believe that family support is especially critical to UMTYMP. Recommendations include discussing classroom activities with children, providing a time and place for them to study, and participating in conferences if concerns arise. It also encourages parents to complement the program's efforts by becoming aware of the breadth of mathematics topics covered in their children's classes and becoming more involved in supporting the good work habits that are critical for success in UMTYMP.

In an effort to sustain parental involvement, UMTYMP has established a variety of academic-year activities for parents and students such as orientation sessions, a workshop on learning styles, and a mathematics fun fair. The academic progress of all students is monitored by an UMTYMP counselor, and there is continued emphasis on regular counseling contacts with the parents.

Social events have become a key component of the intervention project and have been extended from girls-only in the Algebra class to all levels of the program. While the socialization began with female-only events such as bowling parties, movies, and pizza parties, the students have now expressed a strong desire for coed parties, as well as a desire to become a part of the planning process. A letter from a concerned female geometry student suggested that rather than isolate the girls by sponsoring all female parties, we should bring the girls into the mainstream with coeducational functions. She also suggested the establishment of a Student Board to plan the activities. Both of these suggestions are being implemented.

To support the perspective that UMTYMP not only provides a mathematics education but also encourages a culture of mathematics, older students participate in several social events which focus on college and career information. One such event was a calculus luncheon which featured a social hour followed by alumni presentations on the impact of UMTYMP on their college experiences. The presenters included four female alumni, two of whom are Ph.D. candidates, and three male alumni. In an effort to provide women role models and possible mentors for calculus females, a "shadow" program is being established whereby these young women will have the chance to share the experience of women working in various fields of study. The Learning

Styles Workshop and the Math Fun Fair round out the cultural experience of the younger UMTYMP students.

New testing procedures and special orientation meetings for females who were successful in the qualifying examination were established. Activities to encourage and support females who nearly qualified were also included in the intervention process. The near-qualifiers were invited to a variety of social activities with current UMTYMP females. These interventions will continue this year. Inclusion of the near-qualifiers has proved to be very successful in 1989–1990, with a consistently large number of near-qualifiers attending the functions. Moreover, the pool of near-qualifiers in 1988–1989 provided an outstanding source of successful UMTYMP participants for 1989–1990. We hope to see similar success in future years.

A program was developed for selected schools in which key teachers encouraged increased participation of their more mathematically talented female students in UMTYMP. Last year's target school intervention had a slow start and was not as successful in identifying students as we had hoped it would be. Procedures are now in place to improve this intervention project. This year's contact teachers are also being asked to identify a smaller number of females (between five and ten) to make the task more manageable. One of last year's contact teachers has returned to the project and provides insights and guidance to the new teachers. We believe that in the long term these personal contacts will have a positive impact.

With partial support from the National Science Foundation (NSF) Young Scholars Project, a two-week special Summer Institute for current UMTYMP students and a Summer Enrichment Institute for prospective UMTYMP females was again offered in June 1990. Both institutes, held on the University campus, are special enrichment-oriented programs that teach interesting topics different from those in the usual curriculum, while incorporating problem-solving techniques and mathematics league competition skills. The residential Summer Institute is designed for continuing UMTYMP students, while the nonresidential Enrichment Institute is a half-day program designed to increase the interest of female near-qualifiers and Spring qualifying students to participate in UMTYMP. Along with enrichment activities, the institutes provide industrial tours, career information, and socialization activities for students. Outstanding high school teachers from the academic year UMTYMP class, as well as University of Minnesota mathematicians, are involved in the program.

The 1989 Summer Institute attracted twenty-six students, one-third of them female, and the Summer Enrichment Institute consisted of fourteen prospective female students. Evaluation results from the Summer Institute indicated that the institute was a very positive experience for the participants. In fact, a survey indicated that nearly three-quarters of the participants want to return in 1990. The success of the Enrichment Institute on its female participants was reflected in parent comments such as: "My daughter had a positive experience and has a better understanding of the program and her own thoughts about being in UMTYMP," and "We feel this program will increase her confidence in her mathematics abilities and will also motivate her to continue."

There were two special institute activities in 1989. Students who had completed at least one year of calculus were given an opportunity to work with Dr. John Hubbard, a mathematician with special interests on computer applications to geometry, on his innovative educational software. Several world-class mathematicians associated with the Geometry/Supercomputer Project and UMTYMP co-hosted an extraordinary closing event for both institutes, local high school teachers, and other potential students. A luncheon for the students included University of Minnesota President Hasselmo, members of the Geometry/Supercomputer Project, business leaders, and other university mathematicians. This luncheon was preceded by a highly informative talk on the geometric aspects of tilings by Dr. William Thurston, Princeton University, and followed by an exceptional presentation (with videos and films) by the founder of fractals, Bènoit Mandelbrot. His talk on the "Fractal Cosmos" was so alluring that many students asked for his autograph at the conclusion!

In 1990, further improvements were sought for these institutes. Speakers and activities were carefully reviewed as to their appropriateness for students' age-level and mathematical knowledge. The Enrichment Institute had more tours (3M, Physical and Chemical Engineering Departments, and the Supercomputer Institute) and fewer speakers. Last year's programs were expanded and provided increased opportunities and curricula for mathematical enrichment. For example, a major theme of symmetry was introduced via projective geometry. A teaching format with two teachers in each class was utilized. New programs and additional involvement with the Geometry/Supercomputer Project and the Institute for Mathematics and Its Applications (IMA) was sought. Videos developed by the Geometry/Super computer Project were shown to the students.

## Progress Report

Although the Bush Project is just completing its first phase, initial results are very promising. In September 1988, partial implementation of the new testing procedures has already resulted in a much stronger applicant population. Using a more difficult qualifying test and a higher cutoff from previous tests, a class 20% larger than usual was admitted. The testing population was 44% female, and the entering class was 32% female. Despite higher standards, female scores showed an overall improvement and were more evenly distributed than in the past. In the Spring of 1989, an improved preregistration procedure was implemented which involved supplying schools with preregistration packets to be distributed to students. Parents then sent the preregistration form directly to the UMTYMP office. These new procedures created an even stronger applicant pool, and further gains in female enrollment were realized. In 1989, the testing population was 45% female, and the entering class was over 40% female. That was an 83% increase over the 1987–1988 academic year which had no interventions. Furthermore, the females who preregistered as a group in the Spring scored two points higher on the qualification test than the females who decided to register through the schools in the Fall. The statistical difference between the means of the two groups is highly significant. Thus, these new procedures are identifying better qualified female applicants.

Further examination of prior data suggested that the near-qualifier pool should be expanded to include those females who scored within ten points of qualifying for the program. (Previously, near-qualifiers needed to score within four points of qualifying.) This larger pool of the 1988 near-qualifiers was either given an opportunity to take a Spring qualifying exam or mailed a letter urging them to retest in the Fall of 1989. Of those forty-nine who retested, sixteen were accepted into the program. This is a remarkable qualifying rate of 32%, far higher than the historical qualifying rate of 8%–10%. Overall, the new strategies to encourage near-qualifiers and implement more equitable enrollment appear to be quite successful.

Based on the literature and prior UMTYMP experience, it was decided that a class with 50% enrollment of each sex would be most desirable. In 1988, three algebra classrooms were 50%–50% and two were all male. Equal enrollment classes have led to much improved and more supportive classroom dynamics for the female students. In fact, one class was clearly dominated by its strong, vocal female group in the first year. Together with the social opportunities and workshops, this structure has led to improved retention of girls. In 1989–1990, four of the five Algebra classes, and three of the four Geometry/Math Analysis classes are 50%–50%.

In the first year of intervention, the retention rate from Algebra I to Algebra II for males and females was essentially the same (91.9% for females, 92.1% for males). Prior to any interventions (the 1987–1988 school year), the corresponding retention rates were 85.7% for females and 90.7% for males. The retention rates from Algebra II to Geometry in 1988–1989 were nearly identical for males and females (85.3% for females, 85.7% for males). The corresponding rates in 1987–1988 were 83.3% for females and 84% for males. The 1988–1989 retention rates represent a major difference for female students over 1987–1988. The preliminary retention results for 1989–1990 are even more impressive. As of March 1, 1990, in *every* subject in UMTYMP, there have been fewer withdrawals by female students than by male students. Moreover, overall retention is very high, with only 9% withdrawal in Algebra and 7% withdrawal in Geometry/Math Analysis. Most surprisingly, calculus retention for females has dramatically improved. In fact, in the very demanding Calculus III (Linear Analysis) course, all three female students have stayed in the course, while several of the male students have withdrawn.

The three females are a tightly knit unit both academically and socially. It appears that resolution of some of the issues of isolation and lack of significant friendships with other UMTYMP females have a dramatic impact on retention.

The success of these interventions can sometimes be measured in a different context by observing the behavior of the students. Near the end of the 1988–1989 Spring term, the Algebra students were given the opportunity to request classmates for their next year's Geometry class. In the formation of these classes we were able to honor almost every request. One class in which the females dominated the sessions was particularly adamant about being placed in the same Geometry class for the following year. One week into the 1989–1990 school year, students from that class repeatedly called a girl who had dropped out of UMTYMP and tried to convince her to rejoin them. She decided to reenter three weeks into the Fall term. As another example, on the first day of Fall 1989 classes, returning students excitedly greeted friends and rushed to the bulletin board where the class lists were posted to find out with whom they would share their Geometry class. Finally, one of the Geometry teachers had to miss class so her students were distributed into the other three Geometry classes. Her students said that, despite the fact that the other teachers were good, they were anxious to return to their class and learn with their friends.

The overall aim of the Bush Intervention Project is to create a supportive environment in UMTYMP which encourages capable females to seriously involve themselves in mathematics. The success of these interventions to provide this atmosphere is also indicated by the following quotes:

> I realize the importance of the early encouragement I got in UMTYMP; I will pass on that encouragement to other girls as a role model and through volunteer projects. (An UMTYMP graduate and current UMTYMP Teaching Assistant.)

> Last year was our daughter's first and I saw the benefits in light of her growth and development and love of mathematics. (An UMTYM parent.)

> She has never thought of herself as "very good" at mathematics, having been in a class with other equally good students. However, the reality of it is now dawning on her and it has really made a difference to her self-confidence. (An UMTYMPP parent.)

## OTHER ACTIVITIES

Along with the activities already mentioned, plans for this year include a summer employment program. UMTYMP alumni who are currently in college are encouraged to send a resume to the UMTYMP office. In 1989, twenty-eight resumes were received with 21% of these resumes coming from female students. These resumes were sent to nine companies according to the interest of the student and the needs of the companies for summer employment. Partial data indicates that at least five students were placed in jobs. Three of these students worked at the University of Minnesota Supercomputer Center in the Summer of 1989 doing innovative undergraduate research with the Geometry/Supercomputer project. Despite the fact that these students were only freshmen, their participation level and contributions were excellent. The placement program has been expanded this year to include more companies. In Summer 1990, eight students were hired by the Geometry Project for summer research, including two who had worked for the project the previous summer. Their experiences were even more rewarding the second summer, and their research activities were valuable and appreciated by the Project.

Both the UMTYMP graduates and the employing companies are very enthusiastic about the program. Both benefit considerably when creative jobs are available. Moreover, as other alumni have become aware of the summer employment program, demand for these types of opportunities has increased.

The program has always hired alumni attending the University and nearby local colleges as teaching assistants during the academic year. About 50% of the teaching assistants are female. These teaching assistants take on a considerable teaching activity while maintaining a full academic program in highly demanding majors. The first group to have spent three or four years as teaching assistants in the program are now graduating and applying to top graduate schools for Ph.D. programs in science and engineering. We were especially encouraged to note that one of the female teaching assistants has obtained admission and a fellowship to a major Ph.D. graduate program (University of California, Berkeley in Genetics). Moreover, another graduate has sought employment in the area of Chemical Engineering. She obtained several extremely attractive offers and has accepted a position at Proctor-Gamble in product development.

These teaching assistants have been outstanding role models to the UMTYMP students. The program is now beginning to see the effect of their presence. Two of the Calculus III females have applied to be academic teaching assistants during the 1990 Summer Institutes. More of the younger females are showing signs of leadership within the academic year program. We believe that many new female role models for UMTYMP will emerge in the near future.

## CURRICULUM INNOVATION: A NEW TECHNOLOGY-BASED ALTERNATIVE COURSE

The program has always been aware of the situation that many highly capable students might benefit from variations on the standard UMTYMP curriculum. In fact, some of the successful students currently enrolled in UMTYMP found it helpful to repeat portions of a course already completed or possibly withdraw from the program for some portion of the year. A slower-paced period, or the opportunity to enrich their background in a subject, is frequently quite beneficial. Both prior and subsequent to these periods, these students progress without difficulty through the standard UMTYMP courses. Sometimes they return to be top students after this period of change.

Another situation sometimes encountered is having capable and motivated students admitted to the program who subsequently discover some serious gaps in their mathematical skills or some major problems with their study habits. With intense counseling and major extra efforts, these students can progress in the program, but the extra work on top of the already heavy demands of the regular curriculum can be too burdensome. Sometimes these students are forced to withdraw.

As a method to address both of these situations, UMTYMP considered developing an alternative model which would maintain the same standards and basic curriculum as the current two-year high school program, but it would provide a more enriched and more customized program over a three-year period. The somewhat slower-paced program would allow more time to enrich standard topics, provide additional background or an individualized basis when necessary, and somewhat lessen the heavy homework commitment of the regular program. An additional tutorial class each week would encourage group work and appropriate study habits, provide focused activities on high-level problem-solving skills, and generally provide whatever assistance and support is necessary. This model would ultimately allow students in the standard curriculum to transfer to this alternative if the need arose. Conversely, students completing the alternative curriculum should be able to handle the UMTYMP Calculus program as well as the other students. Finally, the flexibility and individualization of the alternative program would hopefully encourage greater participation in UMTYMP by underrepresented groups. The opportunity to work individually with motivated and talented students and to shape some of the skills necessary to succeed in UMTYMP would provide another important dimension of equity.

The recent emergence of graphing calculators in the secondary curriculum provided another direction for this alternative model. With the pending availability of the more user-friendly second generation graphing calculator (e.g., the TI-81) and the general availability of better software, it was decided to emphasize technology in this alternative course. A new curriculum which will totally integrate graphing calculators will be developed. Computers will be used for classroom demonstrations, and the new materials will follow the guidelines suggested by NCTM *Standards*.

Thanks to a recent commitment in March 1990, by the Cray Foundation to support a four-year pilot of this new alternative course, the program will begin to implement this program beginning in 1990-1991. One of the outstanding UMTYMP high-school teachers—a Presidential Awardee nominee who is highly proficient in the use of technology in the classroom—has agreed to develop the curriculum and teach the pilot course for the full three-year period. During 1990-1991, he will prepare the first-year curriculum and seek suitable textbooks and supplementary materials. Also, he will generally design the outline of the three-year program. Based on the scarcity of materials using the graphing calculator currently available, much of this innovative curriculum will have to be locally developed.

With the background and knowledge from the Bush program, UMTYMP will initiate a variety of activities to ensure that the 1991-1992 pilot class will achieve some broader representation among students of color. Special contacts and commitments will be sought from the schools, community organizations, and families. These will include school visits, community visits (including churches and recreational groups), and parent meetings. An advisory group of parents of students already in UMTYMP, community leaders, and key school contact persons will be formed. Most of 1990-1991 will be devoted to gaining credibility in the various communities and identifying a pool of potential students for the alternative pilot. A special Spring test will be administered, and a residential Summer Enrichment Institute will be offered in Summer 1991. This enrichment institute will enable the students to meet one another, become familiar with the instructor and the graphing calculator, begin to understand how UMTYMP functions and its approaches and expectations about curriculum, and, last but not least, study some interesting enrichment mathematics.

UMTYMP is very excited about this alternative course and the potential benefits to its students and to the program in general. We fully expect it to be yet another example of the enriching influences and improved learning environment that the equity and excellence activities have brought to the Talented Youth Program.

## References

1. T. Berger and H. Keynes, *The challenge of educating mathematically talented students: UMTYMP*, CBMS Issues in Mathematics Education (N. Fisher, H. Keynes, and P. Wagreich, eds.), vol. 1, Amer. Math. Soc., Providence, RI, 1990, pp. 11-32.

2. Lynn Steen, ed., *Everybody counts: a report to the nation*, MSEB/NRC, National Academy Press, Washington, DC, 1989.

DEPARTMENT OF MATHEMATICS, UNIVERSITY OF MINNESOTA, MINNEAPOLIS, MINNESOTA 55455

# A Report on an Entry Level Math Program

ELIAS TOUBASSI

## Introduction

In 1985, the Mathematics Department of the University of Arizona began an extensive program to improve the teaching of its lower-division mathematics courses, known as the entry level courses. The Entry Level (EL) Program focused on seven basic courses: Introduction to College Algebra, College Algebra, Trigonometry, Finite Math, Precalculus, Business Calculus, and Calculus I. Depending on their preparation in high school, practically all university students, especially students in the Colleges of Arts and Sciences, Business, and Engineering, are required to take one or more of these courses. Thus, student success in these courses enables them to continue their course of study; whereas, failure in these courses effectively closes many academic options.

The EL Program, under the direction of Professor Elias Toubassi, with the assistance of the Entry Level Committee consisting of Professors William E. Conway, Richard S. Pierce, and Frederick W. Stevenson, was conceived as a comprehensive program and built on broad faculty support. Key elements of the program are instituting a mandatory math placement test, providing a positive learning environment, recruiting a high quality adjunct teaching staff, and developing contacts with the local schools. The EL Program has been dramatically successful in increasing the enrollment and pass rate in the lower-division mathematics courses, and its impact has been felt throughout the university and the local school districts.

In this paper, we describe the circumstances leading to the implementation of the EL Program, descriptions of the program's key elements, and data about student enrollment and achievement. A concluding list of "essential ingredients" is offered for consideration by other mathematicians who may be interested in implementing similar reforms within their own departments.

**Background: Circumstances leading to establishing the Entry Level Program.** During the 1970s, the University of Arizona, like many other colleges and universities, experienced a large growth in enrollment in what is known as the

Entry Level mathematics courses: Introduction to College Algebra, College Algebra, Trigonometry, Finite Math, Precalculus, Business Calculus, and Calculus I. From 1969–1976 enrollment in these courses increased by 67% from 6,538 to 10,915. At the same time that student enrollment was increasing, administrative decisions led to a change in the composition of the teaching staff. The number of faculty positions was reduced, while the number of teaching assistant positions was increased. The situation forced the mathematics department to make some difficult choices in order to meet its teaching obligations: thousands of students in Intermediate and College Algebra went into a self-study learning program, and thousands more students in Finite Math and Business Calculus were taught in lectures of size up to 600. Students and teachers were frustrated by the depersonalized styles of instruction, and attrition rates (drops plus failures) were very high, and over 50% for some courses.

The department's calls for additional faculty and operating budget went unheeded for a dozen years. The first signs of change appeared in the early 1980s with the administration of President Henry Koffler, Provost Nils Hasselmo, and Ed McCullough, Dean of the Faculty of Science. (Support has continued under the current administration of Provost Jack R. Cole.) One of Provost Hasselmo's first initiatives on coming to the University of Arizona was to appoint a university-wide committee to look into freshman mathematics. The nine member committee, chaired by Vern Johnson, Professor of Electrical and Computer Engineering, was appointed in the Spring of 1983 and consisted of the Dean of the Faculty of Science and faculty members from mathematics, engineering, business, and social science. In its report a year later, the Committee expressed concern over the high drop/failure rates and the low grades earned in Entry Level mathematics courses. The Committee identified Intermediate Algebra and first semester Calculus as having "fundamental problems." During a five-semester period from 1981–1983, only 48% of the Intermediate Algebra students received a passing grade (A through D) while in Calculus I only 44% received a passing grade.

The Committee's investigation included a study of the resources and enrollment figures in the Department of English as compared to the Department of Mathematics. Striking imbalances were revealed. In the Fall of 1983 Entry Level math courses had an enrollment of about 6500 students and the department had a budget to support forty-seven faculty and ten full-time equivalent (FTE) graduate assistants. In contrast, Entry Level English courses had 4100 students and the department had a budget to support sixty-three faculty and thirty-three FTE graduate assistants. The Committee concluded that: "Although other factors need to be considered, it is difficult to avoid the conclusion that many of the problems in freshman math can be traced to inadequate allocation of resources by the University Administration."

Beyond recognizing the need to allocate additional resources for teaching staff in mathematics, the Committee made several far-reaching recommendations, which are discussed below. Two committees were appointed: an Ad Hoc Intercollegiate Committee on Calculus and a Departmental Entry Level Committee. The Committee on Calculus was charged to revise the entire calculus sequence and propose suitable recommendations. The Entry Level Committee was charged to draft a plan of action to address the problems related to Entry Level mathematics courses. In its report in May 1985, the Entry Level Committee outlined a five-year implementation plan and the resources needed for each year of the plan. The report also included several studies that compared the mathematics department with other departments within the University and with mathematics departments at other universities, including the mathematics departments at the Universities of Illinois at Urbana, The Ohio State University, and the University of Wisconsin, Madison. All of the studies suggested that the mathematics department at the University of Arizona was doing more teaching with fewer resources. The report recommended the creation of a comprehensive training program for teaching assistants (TA), a feasibility study on the use of computers to aid instruction, more office space for new faculty and TA's, and more classrooms for instruction.

**Description of the EL Program.** The EL Program was founded on four fundamental premises:

(1) Students must be placed in courses commensurate with their abilities and mathematical background.
(2) Students must be provided with a supportive learning environment and taught by caring instructors, who are committed to undergraduate education.
(3) EL courses must be structured to meet current student needs.
(4) The future success of the program relies on an effective outreach to the schools.

In addition, it was felt that students must be able to use modern computer technology.

To describe the way in which the plan was implemented, the recommendations of the department's Entry Level Committee are listed, followed by the details of how and when they were implemented.

# RECOMMENDATION 1:
## INSTITUTE A MANDATORY MATH PLACEMENT TEST

Although there was an advisory Math Readiness Test (MRT) in use at the time of the EL Committee report, the testing program was not working well as a placement tool. Twenty-five to fifty percent of entering students did not take the MRT, and the majority of the students who did take the test ignored the results. A change was made in the Spring of 1986 when all students enrolling in EL courses were required to take the MRT. Mandatory placement was implemented beginning with Calculus, in the Spring of 1987, and completed in the Fall of 1988 with the inclusion of the algebra courses. Since then, minimum requirements have been set for all the EL courses, except for the lowest level algebra course. Students' MRT scores and their cumulative high school grade point averages (GPA) are used for evaluating their mathematics preparation for course placement. The math readiness testing program is directed by Associate Professor Ted Laetsch with the assistance of Rick Kroc, Director, and Reed Mencke, Associate Director, from the University Testing Center.

## Recommendation 2: Introduce a Calculus I Course with Five Semester Credits

The EL Committee believed that the three one-hour lectures per week were insufficient for the majority of Calculus I students to adequately learn the fundamental concepts of calculus. This belief was supported by a survey of a dozen comparable universities. The average calculus sequence at the surveyed schools was 12.6 semester credits, compared to Arizona's 10 credits. In the Fall of 1986, the department introduced a five credit course in Calculus I alongside the existing three credit course. (Experimental sections of this course were introduced in the Fall of 1983 by Associate Professor Richard Thompson.) The two additional meetings per week permit a slower pace with more time for exercises and applications. The requirements for entry into the three-credit calculus course are very stringent. Only 20–25% of the Calculus I students qualify for it. This arrangement has worked extremely well during the past four years. The department has also continued its effort to improve the calculus curriculum. (See p. 111) for a brief discussion of curriculum changes.)

## Recommendation 3: Institute a Precalculus Course with Three Semester Credits

It was felt that aspiring science and engineering students who do not meet our placement requirements for Calculus I need a suitable course to prepare them for success in calculus. A precalculus course to review and strengthen the basic skills in algebra, trigonometry, and analytic geometry was introduced in the Spring of 1987. In a survey of precalculus students who went on to Calculus I, most felt the course fulfilled its intended goal. Further, when Calculus I instructors were asked to compare the performance of precalculus students with other students, all of the instructors felt that the two groups were roughly comparable. Associate Professor William E. Conway, the architect of this course, was partially supported by a Provost Creative Teaching award during the time he developed the course.

## RECOMMENDATION 4[1]: REPLACE THE INTERMEDIATE AND COLLEGE ALGEBRA COURSES WITH A TWO-SEMESTER SEQUENCE IN COLLEGE ALGEBRA TO BE TAUGHT PRIMARILY IN SMALL LECTURE CLASSES

Our previous algebra program suffered from two problems: an outdated curriculum and an unsatisfactory teaching format. The ad hoc committee analyzing the algebra curriculum found there was too much duplication in subject matter between the one-semester three-credit Intermediate Algebra course and the one-semester three-credit College Algebra course. Moreover, the university's new admission standards, which required entering students to have completed two years of algebra and one year of geometry, took effect in the Fall of 1989. A two-semester sequence in College Algebra, which provides a more comprehensive study of the subject and thus better prepares students for further work, was introduced. It was also recommended that the algebra courses be offered in lectures with thirty-five students as an alternative to the self-study teaching format which was the predominant learning scheme at the time. Although self-study is effective for some students, especially older, self-motivated students, the self-study program had not been working well for the majority of students and had acquired a bad reputation. It was decided to try to offer as many lecture sections as possible, subject to availability of faculty. During the 1989–1990 school year, the department came close to achieving its goal of giving all students a choice between self-study and lecture classes. Adjunct Lecturer Marlene Hubbard is the coordinator of the algebra program.

## RECOMMENDATION 5: REDUCE CLASS SIZE IN FINITE MATH AND BUSINESS CALCULUS

One of the goals of the EL program is to move away from the depersonalized large lecture classes of 300–600 students, which have relatively low pass rates, to more successful small lecture classes of thirty-five students. We have not yet attained the goal of providing small lecture classes to all the students who want to take them. In Finite Math, about two-thirds of the students are in small classes. In Business Calculus, the 600-student lecture class has been divided into seven lecture classes of size around ninety each. Our plan is to continue to reduce class sizes until we reach our target of thirty-five students per class.

---

[1] This recommendation is slightly different from the one in the 1985 EL Committee report. At that time it was proposed that Intermediate and College Algebra be replaced by a one-semester course in College Algebra.

In addition to pursuing the recommendations discussed above, the mathematics department has adopted policies to improve student-faculty relations. These policies include recruiting caring classroom teachers, developing outreach programs to the schools, and introducing technology in instruction.

**Caring classroom teachers**. It is our intent that every student study mathematics under the guidance of a skilled, caring teacher. Fortunately, we have been able to augment our regular faculty and teaching assistant staff with a dedicated group of adjunct faculty who are committed to undergraduate education. In 1989–1990, forty-two part- or full-time adjunct faculty equivalent to 24.25 FTE were hired. The adjunct faculty, who are enthusiastic about working with beginning undergraduate students, are teachers who are recruited from local schools or participants in the University-School Cooperative Program. (See below for a disussion of the Co-Op program.) In addition, the university and the department have introduced a comprehensive training program for new TAs. The new TAs are required to attend an orientation session conducted by the Graduate College, followed by a three-day departmental training program, which includes such items as lecturing techniques, use of blackboard, test construction, and student faculty contacts. Follow-up meetings are also held during the academic year. The departmental component of the TA training program is directed by Associate Professor Dan Madden. Non-native TAs, who comprise 40% of the TA staff, have to fulfill further requirements before they can become classroom teachers. They attend a culture pedagogic workshop offered by the Center for English as a Second Language and must pass the Test of English as a Foreign Language (TOEFL) and the Test of Spoken English (TSE).

**Outreach to the schools**. An important goal of the EL program is to develop bridges to the schools that will strengthen the educational program of both institutions. We felt that university students in EL courses, as well as prospective mathematics teachers, would benefit from interaction with skilled school teachers. Conversely, we believe that junior high and high school students are likely to have a better understanding of the expectations in university mathematics courses if their teachers have first-hand experiences at the university. Our links with the schools take several forms. The new University-School Cooperative Program, started in 1987, enables teachers to spend a year on campus,while receiving full pay from their district, to teach and take courses, to participate in a mathematics instruction colloquium, and to work on a project of their choice. Professor Stephen Willoughby directs the mathematics instructional colloquium. The program has been extremely successful in providing teachers with opportunities for professional growth and opening up dialogue between teachers and university faculty, and we are most excited about its future promise. To bolster the program's impact, we offer courtesy appointments to the teachers in the Co-Op program after they return to their home school so that they can continue their collaborative work with the mathematics department on an ongoing basis. There are also teachers with quarter-time positions in the EL program, who teach late afternoon or early evening classes. Thus far, about three dozen teachers have participated in some phase of the EL program. A third part of our outreach program is the Presidential Fellowship Program directed by Professor Art Steinbrenner. In this program, teachers receive support which allows them to come on campus to join the graduate program and work toward a master's degree.

Other programs between the schools and the mathematics department include a summer math camp for high ability junior high students initiated by Associate Professors Dan Madden and Fred Stevenson, and two National Science Foundation (NSF) teacher enhancement projects aimed at integrating problem solving and technology into the curriculum. One project, directed by Associate Professor David Gay and Pima College Assistant Professor Debbie Yoklic, is for junior high mathematics teachers in Arizona; the other project, directed by Professor Elias Toubassi, is for junior high and high school mathematics teachers in the Tucson area.

**Technology in instruction.** Another instructional innovation is the use of technology to support and enhance learning. In 1986, Professor Clark Benson piloted the use of algebra homework on computers. Currently, algebra students are required to do seven or eight units of computer homework to sharpen their skills. We have found that over 88% of the students who complete five units or more of their computer homework earn a course grade of A, B, or C as compared to 70% of all students who earn these grades. In addition, an "Are You Ready" disk was created and distributed widely. Its purpose was to alert students to the algebra skills needed to succeed in Calculus I. Other "Are You Ready" disks have been completed for Calculus II, III, and Ordinary Differential Equations. These efforts are continuing with the creation of an undergraduate electronic classroom, which has thirty microcomputers, one per student. Bob Condon, System Manager of the Computer Lab, and Professor David Lovelock have played a key role in planning and designing the electronic classroom, creating software packages, and training faculty in the use of the software.

## COMPARATIVE DATA AND RESULTS

The primary evaluation of the EL program centers on two measures identified by the University Committee on Freshman Mathematics: (1) the pass rates in these courses and (2) the grade point averages earned.

In addition, anecdotal information has been gathered from instructors, students, and university faculty on their reaction to the program. The first comparison examines the difference in performance between students who took EL courses in the six semesters of Fall 1982 to Spring 1985, prior to the existence of the program, and those who took the same courses, in small lecture classes, under the new EL program in 1985–1989. Figure 1 shows that the new EL sections had overall pass rates which are 42% higher than pre-EL sections and earned grade point averages, which are significantly higher. The course-by-course breakdowns follow in Table 1.

It should be noted that three courses, Trigonometry Math 118, Precalculus Math 120, and Business Calculus Math 123 are not in Table 1. Math 118 has always been taught in small lecture classes, and for the past five years it has been used as a TA training course; Math 120 is a new course started in 1987; Math 123 has undergone some reduction in class size and more

PASS RATE

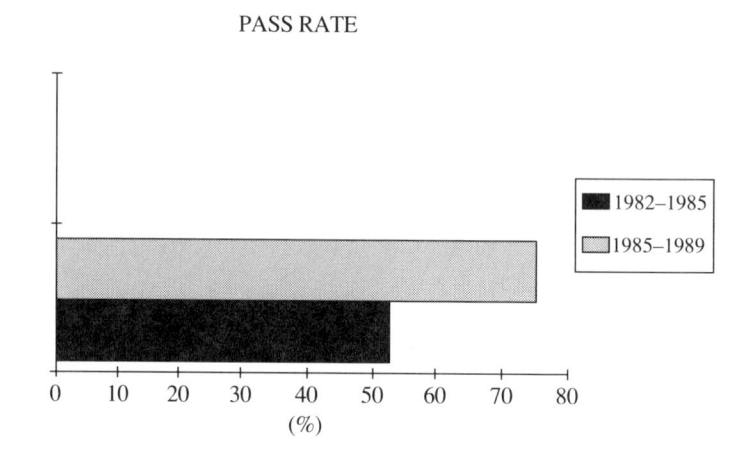

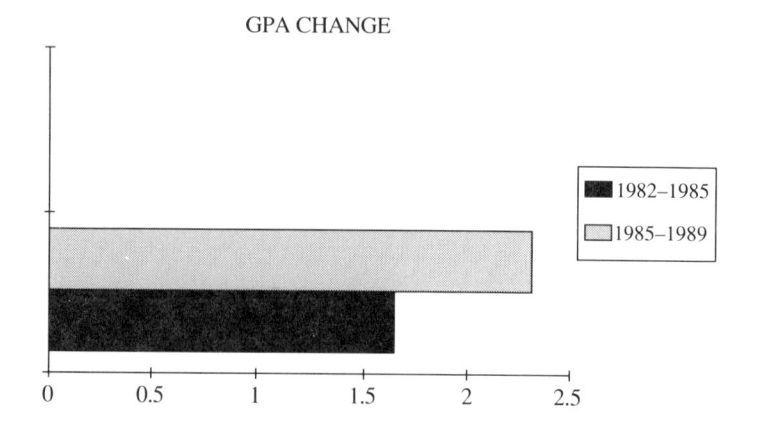

GPA CHANGE

FIGURE 1. Comparison of the overall pass rate and GPA [2] of new sections with pre-EL sections.

---

[2] The GPA is based on the following scale: $A = 4$, $B = 3$, $C = 2$, $D = 1$, $E = 0$.

TABLE 1. Compares pass rates and GPA
of new sections with pre-EL sections.[3]

| Course | Teaching Format | A–D | GPA |
|---|---|---|---|
| Math 116 | 1982–1985 (6 semesters) pre-EL [4] | 44% | 1.26 |
| Intro. to Coll. Alg. | 1986–1989 (7 semesters) new sections | 74% | 2.21 |
|  | CHANGE | +68% | +.95 |
| Math 117 | 1982–1985 (6 semesters) pre-EL [4] | 56% | 1.74 |
| College Algebra | 1985–1989 (9 semesters) new sections | 76% | 2.32 |
|  | CHANGE | +36% | +.58 |
| Math 119 | 1982–1985 (6 semesters) pre-EL [5] | 63% | 1.73 |
| Finite Math | 1986–1989 (7 semesters) new sections | 78% | 2.39 |
|  | CHANGE | +24% | +.66 |
| Math 124/125A | 1982–1985 (6 semesters) pre-EL [6] | 47% | 1.91 |
| Calculus I | 1986–1989 (7 semesters) new sections | 72% | 2.38 |
|  | CHANGE | +53% | +.47 |

reductions are planned (see the discussion under Recommendation 5). Some data on these courses can be found in Table 2 (see p. 106).

More dramatic changes have taken place during the last two years, when the funding level has been substantially increased. Figure 2 (see p. 106) compares the overall pass rate and GPA of the pre-EL base years, 1982–1985, with the last two years of the program. The overall pass rate was 50% higher, and the GPA has increased by almost a full grade point. The course-by-course breakdowns for this comparison follow in Table 2. What is remarkable about this data is that it includes all EL courses irrespective of teaching format. In addition to the new, small lecture classes, it contains data from self-study and large lecture classes. The latter teaching formats have done rather well in recent years due to various initiatives. Due to the substantial reduction in the enrollment in the self-study algebra program, it is possible to conduct an extensive advising program which is supported by an algebra tutoring service. The success of the large lecture classes in Finite Math and Business Calculus are due mainly to the efforts of a few regular and entry-level faculty who have the ability to handle the large lectures.

Table 3 (see p. 107), contains a grade distribution and GPA for all EL courses year by year, regardless of teaching format, starting with the pre-EL years 1982–1985 all the way to the present. It provides a good historical perspective of the changes that have taken place.

---

[3] The source for this and other grade data comes from the Grade and Unit Summary Reports from the Registrar's Office.

[4] Primarily a self-study teaching format.

[5] Primarily lecture classes of size 330.

[6] Taught in three-credit format and in classes primarily of size forty to fifty.

TABLE 2. Compares pass rates and GPA
of pre-EL years with last two years.

| Course | Teaching Format | A–D | GPA |
|---|---|---|---|
| Math 116 Intro. to Coll. Alg. | 1982–1985 (6 semesters) pre-EL | 44% | 1.26 |
| | 1988–1989 (3 semesters) | 78% | 2.32 |
| | CHANGE | +77% | +1.06 |
| Math 117 College Algebra | 1982–1985 (6 semesters) pre-EL | 56% | 1.74 |
| | 1988–1989 (3 semesters) | 85% | 2.44 |
| | CHANGE | +52% | +.70 |
| Math 118 Trigonometry | 1982–1985 (6 semesters) pre-EL | 55% | 1.97 |
| | 1988–1989 (3 semesters) | 71% | 2.28 |
| | CHANGE | +29% | +.31 |
| Math 119 Finite Math | 1982–1985 (6 semesters) pre-EL | 63% | 1.73 |
| | 1988–1989 (3 semesters) | 86% | 2.52 |
| | CHANGE | +37% | +.79 |
| Math 123 Business Calculus | 1982–1985 (6 semesters) pre-EL | 61% | 1.87 |
| | 1988–1989 (3 semesters) | 76% | 2.36 |
| | CHANGE | +25% | +.49 |
| Math 124/125A Calculus I | 1982–1985 (6 semesters) pre-EL | 47% | 1.91 |
| | 1988–1989 (3 semesters) | 78% | 2.49 |
| | CHANGE | +66% | +.58 |

PASS RATE

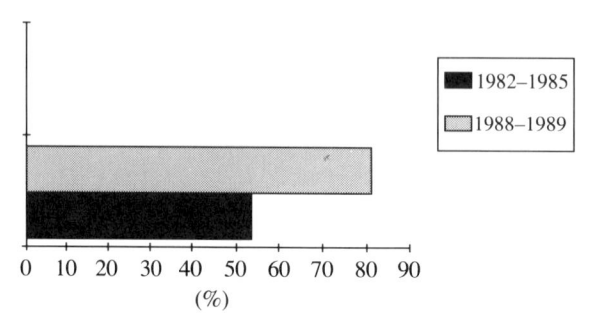

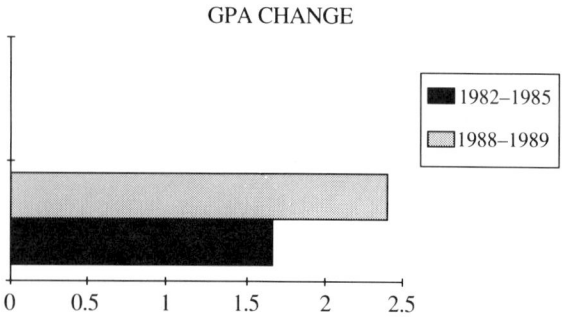

GPA CHANGE

FIGURE 2. Comparison of the overall pass rate and GPA of pre-EL years with the last two years.

TABLE 3. Annual grade distribution and GPA
for all entry-level courses math 116–125A.

| Year | A | B | C | D | E | W | Other | TOTAL |
|---|---|---|---|---|---|---|---|---|
| 1982–1983 | 892 | 1520 | 1788 | 971 | 1673 | 1585 | 257 | 8686 |
| 1983–1984 | 719 | 1197 | 1503 | 1152 | 1852 | 1652 | 275 | 8350 |
| 1984–1985 | 587 | 1130 | 1791 | 1151 | 1723 | 3046 | 270 | 9698 |
| 1985–1986 | 453 | 961 | 1324 | 1028 | 1060 | 1994 | 76 | 6896 |
| 1986–1987 | 1134 | 1735 | 1936 | 1457 | 1495 | 2478 | 270 | 10505 |
| 1987–1988 | 1283 | 2266 | 2395 | 1306 | 1109 | 1906 | 206 | 10471 |
| 1988–1989 | 1733 | 2843 | 2353 | 1136 | 821 | 978 | 160 | 10024 |
| F 1989* | 1398 | 2113 | 1744 | 784 | 661 | 594 | 154 | 7448 |

| Year | %A–D | %E | %W | %Other | GPA |
|---|---|---|---|---|---|
| 1982–1983 | 59.53 | 19.26 | 18.25 | 2.96 | 1.85 |
| 1983–1984 | 54.74 | 22.18 | 19.78 | 3.29 | 1.65 |
| 1984–1985 | 48.04 | 17.77 | 31.41 | 2.78 | 1.64 |
| 1985–1986 | 54.61 | 15.37 | 28.92 | 1.10 | 1.73 |
| 1986–1987 | 59.61 | 14.23 | 23.59 | 2.57 | 1.94 |
| 1987–1988 | 69.24 | 10.59 | 18.20 | 1.97 | 2.15 |
| 1988–1989 | 80.46 | 8.19 | 9.76 | 1.60 | 2.39 |
| F 1989* | 81.08 | 8.87 | 7.98 | 2.07 | 2.41 |

* Fall semester only.

After reading Tables 1, 2, and 3, one might wonder if the higher pass rates are due to easier grading. This is not the case. On the contrary, the grading may have gotten tougher. In the past five years, forty-three of our regular faculty have taught in the EL program. Many of these faculty have taught EL courses prior to the existence of the program, which gives them a historical perspective on the standards of the department. Moreover, entry-level faculty are expected to maintain firm standards. Nowhere is this more evident than in the algebra program where we have set an absolute grading scale (90–100% for an A, 80–89% for a B, etc.) which is adhered to fairly rigorously. The adherence to these standards is supported by evidence that performance in sequel courses has been equal, if not higher, than in the past. We have found that about 75% of the students, who pass Calculus I, continue to complete the sequence of Calculus II, Calculus III, and Differential Equations.

**The EL program and the University of Arizona.** It is important to put the EL program in the context of a growing research university. Faculty at the University of Arizona are hired for their research and teaching potential. Regular faculty have an average teaching load of two courses per semester. The fifty-eight budgeted faculty in the mathematics department cover about 110 of the 279 courses taught each semester. This leaves 169 classes to be taught by TAs and adjunct faculty. Most faculty prefer to teach courses at the level of Calculus I or above. Therefore, a very special program is necessary to care for students in the remaining classes. It must be caring and personalized, and yet it must set high standards.

The University of Arizona has tackled the challenge in what we feel is a wholly unique manner. Additional resources have allowed the department to augment its faculty with adjunct lecturers who are dedicated to teaching. There are several important consequences. First, the number of students who had to take the self-study algebra program decreased in stages. Currently, the annual enrollment is around 1,758 compared to an average of 5,000 in previous years. Second, the mandatory math placement test improved the teaching atmosphere. Class composition became more homogeneous, and faculty could count on the students' having better mathematical preparation. Furthermore, with the addition of adjunct faculty, regular faculty teach the courses that they prefer, and, generally, class size is smaller. The improved teaching conditions have led to increased faculty morale in the department.

The impact on students was even greater although it took them a year or two to believe that the EL program was here to stay. There were noticeable changes from their point of view: they had instructors in their algebra classes, class sizes were smaller, and faculty and staff had a positive attitude toward them and would take time to address their concerns. Less visible to them, but equally important, was the impact of the mandatory math placement test and of the various curriculum changes. These innovations combined to create a reputation for the department as one that cares about teaching and about students. Our changes began to generate positive comments from colleagues across campus and from the Office of the Dean of Students. Our faculty also noticed the difference when their students were telling them more frequently that they enjoyed their math classes. The Dean of the Faculty of Science and the Provost reported to us that not only were complaints about EL mathematics courses dwindling, but that positive things were said to them. One of the nicest things happened when the President of the student body asked to meet with me to tell me of the great things he had been hearing about the EL program. He wanted to know how we accomplished all of this; he had hopes of getting similar programs started in other departments.

The success of the EL program in the past five years exceeded our expectations. My personal goal was to try to bring the drop/failure rates in EL courses down to around 25–30%. If someone had suggested at the time that in five years we would bring those rates down to an average of 19% as Figure 2 shows, I would have called him foolishly optimistic. Morale in the program is high, and there is increasing recognition for faculty who work on matters affecting educational issues. The periodic self-analysis and reporting of results to the Department, University, and Board of Regents has helped keep us vigilant.

A final benefit is the impact of the EL program on students and teachers in the Tucson area. This benefit is likely to pay dividends in the long run. The program has increased teachers' awareness of our curriculum and expectations, increased the dialogue between us through such activities as the Math Instruction Colloquium and the Co-Op Program, revitalized the teachers through the ongoing NSF teacher enhancement projects, Making Math Count and Promoting Reasoning in School Mathematics (PRISM), and it better prepared some high school students through the Junior High Math Camp Program. All of these are indicators of a healthy relationship between the university and local school districts. These collaborative ventures bode well for the future.

## BUDGET AND COST ANALYSIS

The EL budget has grown significantly in the past five years. In the Fall of 1985, we started a pilot program for 225 students with a budget of $50,600; while in the Fall of 1989, we had 8,614 students and a budget of $606,800. Table 4 details the supplemental funding for the EL program from 1985–1990.

TABLE 4. Supplemental funding for entry-level instruction 1985–1990.[7]

|  | Adj inst (inc. Co-Op) | Wages staff | Wages fac | Wages student | Operations | Travel | Capital | Total |
|---|---|---|---|---|---|---|---|---|
| 1985–1986 | $15,000 | $19,100 | $5,000 | $10,000 | — | $1,500 | — | $50,600 |
| 1986–1987 | $100,000 | $16,600 | $16,696 | $10,480 | $15,000 | $1,500 | $35,000 | $195,276 |
| 1987–1988 | $281,500 | $16,500 | $18,500 | $10,480 | $15,000 | $1,500 | — | $343,480 |
| 1988–1989 | $413,310 | $16,500 | $18,500 | $10,480 | $15,000 | $1,500 | — | $475,290 |
| 1989–1990 | $535,800 | $16,500 | $18,500 | $12,500 | $22,000 | $1,500 | — | $606,800 |

Figure 3 (see p. 110) shows the changes in the biggest component of the EL budget, namely, the funding of adjunct faculty.

It should be noted that the annual supplemental funding of the EL program per student is quite minimal. In fact in the 1989–1990 school year there were 15,347 EL students with a supplemental budget of $606,800. This is equivalent to $39 per student. This is roughly 0.5% of $7,174,[8] the annual cost of educating an in-state student. (The cost of educating an out-of-state student is $11,296[8].) This small increase translates into a 50% average increase in instructional improvement, see Figure 2.

What does a 50% instructional improvement for 15,374 students mean? It means there are over 4,000 students (81% of 15,374 less 54% of 15,374, see Figure 2) on campus feeling good about their math skills; it means there are over 4,000 students whose graduation date is not being delayed because of mathematics; and it also means that there are over 4,000 students who do not have to retake their math courses. In terms of dollars and cents this means we do not have to reteach about 118 sections requiring 29.5 FTE faculty at an approximate cost of $295,000. It is a far better proposition to teach students well the first time around.

---

[7] The table was prepared by Faye Villalobos, the business manager of the department.

[8] This information was provided by Mr. Tweet in the university Budget Office.

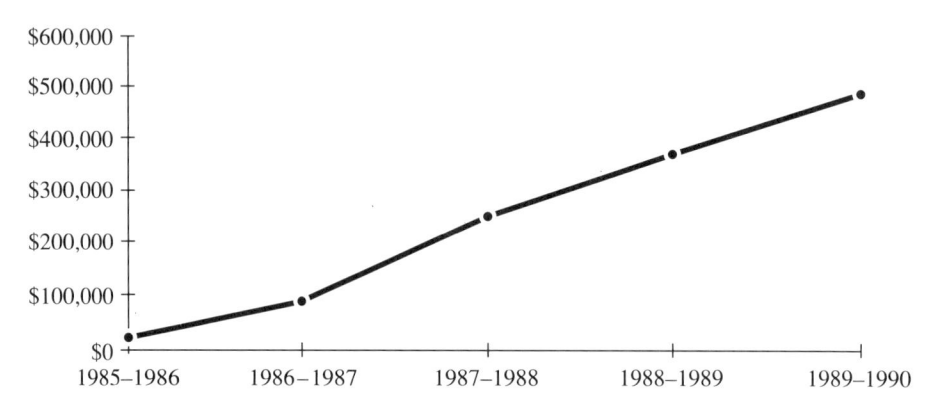

FIGURE 3. Supplemental funding for entry-level instruction 1985–1990.

## THE TASK AHEAD

The next few years will be transition years for the EL program. We will be working to complete the last two recommendations of the original plan and tackling the substantive issues of curriculum reform and the use of technology. A small start on the latter project has been made, but much more needs to be done. With respect to the original recommendations, we would like to have all students taught in small lecture classes. Table 5 shows that there are still over 3600 students who are not in small lecture classes.

TABLE 5. Breakdown of 1989–1990 enrollment by teaching format.[9]

|  | # of students in small lecture classes | # of students in other teaching formats[10] | Total # of students |
|---|---|---|---|
| Fall 1989 | 6415 | 2199 | 8614 |
| Spring 1990 | 5344 | 1416 | 6760 |
| 1989–1990 Total | 11759 | 3615 | 15374 |

Table 6 shows the enrollment in the EL program for the past five years. It is intended to show how far we have come and how far we still have to go. At this time, there is no obvious explanation for the substantial increase in enrollment in the past year. (The freshman and transfer class of 1989 was only 561 larger than the comparable class of 1988.)

For the future direction of the EL program, we are concerned about curriculum reform and using technology to enhance mathematical learning.

---

[9] The source for this and other enrollment data is the Course Summary Reports from the Office of Student Information.

[10] Self-study or large lectures.

TABLE 6. Annual course-by-course enrollment
in the EL program 1985–1990.

| Year/Course | 116 | 117 | 118 | 119 | 120 | 123 | 124 | 125A | Total |
|---|---|---|---|---|---|---|---|---|---|
| 1985–1986 | 1893 | 3000 | 636 | 1574 | NA | 1134 | NA | 1702 | 9939 |
| 1986–1987 | 2011 | 3380 | 749 | 1694 | 88 | 1222 | 1189 | 329 | 10662 |
| 1987–1988 | 2159 | 3919 | 780 | 1877 | 352 | 1167 | 1228 | 381 | 11863 |
| 1988–1989 | 2852 | 3626 | 650 | 1862 | 488 | 1027 | 1119 | 292 | 11916 |
| 1989–1990 | 3339 | 4804 | 891 | 2239 | 717 | 1586 | 1476 | 322 | 15374 |

Many of our current EL courses emphasize skills and computation. We would like to move toward more reasoning and exploration. We would like to integrate, where possible, problems from within the student's realm of interest including such items as data from sports, economics, census, and the like. We would also like to show students that mathematics is relevant and propose problems that they would want to investigate. In our Introduction to College Algebra course, we have started to introduce relevant problems that require problem-solving skills. As we get a clearer picture of the innovations that are most useful, we will incorporate the changes into the curriculum.

The introduction of technology into mathematics courses is proceeding from upper-division courses to EL courses. In the upper-division courses, ordinary differential equations and linear algebra, the use of computers allows the investigation of a larger class of problems and the exploration of graphic solutions to problems. In beginning calculus, several projects are being tried in the 1989–1990 school year. Professor Clark Benson and Adjunct Lecturer Mickey Hoffman are piloting the use of computer labs in the five-credit Calculus course; students are required to do six computer projects during the semester. The labs were written by Professor Clark Benson under a Provost Creative Teaching award. This Spring, Associate Professor Richard Thompson is offering an experimental section of calculus with computers. Next year Professors David Lomen and David Lovelock and Assistant Professor Bill McCallum will try some ideas based on their work in two NSF calculus grants. The first is a University of Arizona grant by Lomen and Lovelock. The second is a multi-year College Consortium Calculus Project. We hope to introduce similar activities in other EL courses over the next few years.

## CONCLUDING REMARKS

In looking back over the work of implementing the EL Program, there are important organizational and policy decisions that have been critical to the success of the program. The concluding list attempts to encapsulate the "essential ingredients" for those who would like to benefit from our experience.

1. Broad faculty involvement.
2. General departmental awareness and support.

3. University encouragement and support for faculty involved in EL activities.
4. Regular communication with interested university departments.
5. A realistic plan of action.
6. Regular contact with student leaders.
7. A mandatory math placement test.
8. Careful teaching assignments.
9. Pride in working in EL activities.
10. Augmentation of regular faculty with instructors dedicated to teaching, including local teachers.
11. Open dialogue and development of joint programs with local school districts.
12. Assessment of the curriculum and definition of the goals for each course.
13. Acceptance of adjunct faculty as colleagues and maintenance of regular communications with them.
14. Development of an effective TA training program.
15. Reward for faculty who make significant contributions to EL.
16. Adequate staff support.

DEPARTMENT OF MATHEMATICS, UNIVERSITY OF ARIZONA, TUCSON, ARIZONA 85721

# Issues and Reactions

# Pros and Cons of Teaching Mathematics Via a Problem-Solving Approach

## BERT FRISTEDT

Since virtually everyone who teaches mathematics at any level requires his or her students to solve problems, the title of this article seems to be lacking in content. However, from various readings and conversations one begins to suspect that more is entailed in the term "problem-solving approach" than would be apparent from the words. My impression is that when people use the term "problem-solving approach" they distinguish "problems" from "exercises," the latter denoting rather routine, drill-type problems. I believe there to be other connotations of the term "problem-solving approach" when used by some people. Some mean that the focus of a particular course should be on "problems" that are inherently interesting and that the relevant mathematics should be introduced as needed. Some also mean that students work together on problems. Finally, for some, the problems in a good "problem-solving approach" should be problems with applications outside of mathematics. Let me use short expressions for the four characteristics I have described: (1) problems rather than exercises; (2) problems rather than systematic development; (3) group learning; and (4) applications.

I am in favor of encouraging group learning, but I think this issue has little to do with a problem-solving approach. Whatever one's overall approach, there are places where group learning can be rather naturally used and places where individual initiative is more appropriate.

Applications generate pedagogical problems whatever method of teaching mathematics one is using. Typically, the more realisitic the application, the more nonmathematical material there is that must be learned.

I will focus most of the remainder of this article on the first two issues described earlier: problems rather than exercises and problems rather than systematic development.

The case I make for problem-solving over exercises goes along the following lines. It is important that certain skills become more or less automatic to students, and long lists of similar exercises are designed to give practice with these skills. I maintain, however, that such long lists of exercises do not necessarily do the job. One reason is that they can be boring, even for a student who is having trouble mastering the skills. The other is that they can convey the message to the student, who is having trouble, that his or her problem is that he or she has not memorized enough different techniques rather than that he or she has not mastered one or two techniques in sufficiently general terms to apply to a variety of exercises. A problem-solving approach in which the need for various skills arises in the middle of doing interesting problems can eliminate or reduce the factor of boredom, and the setting may help the student to realize that the strategy of memorization of every variation that might arise will not work.

I find it easy to convince myself that there has to be a better way to learn than solving long lists of similar exercises. I find it more difficult to decide how to handle problem-solving in such a way that I do not create unforeseen learning difficulties that were not there with long lists of exercises. Thus, I do not feel that one should completely do away with drill exercises.

For an example for discussion, let us consider the circle of ideas connecting Pythagorean triples, monotonicity of functions, and rational numbers. What a pretty story that can be told without calculus—the rational points on the quarter unit circle in the open first quadrant on the one hand correspond in a two-to-one manner with primitive Pythagorean triples and on the other hand are placed into one-to-one correspondence with the rational numbers in the open interval $(0, 1)$ via the function $F$ defined by

$$F(t) = \left(\frac{1-t^2}{1+t^2}, \frac{2t}{1+t^2}\right).$$

The proof can be broken into small pieces. It is an exercise in algebra to prove that the image of $F$ is a subset of the quarter circle of interest, although if one states this exercise with the type of language used here, one has increased the difficulty over what it would be were less concise language used. It is also on the easy side to show that $F(t)$ is a rational point if $t$ is rational. More difficult is the argument for one-to-oneness via a proof of monotonicity and the proof that rational points on the unit circle correspond to rational $t$. A student can be aided by intervention on the part of the teacher or by the breaking of more difficult aspects of the problem into manageable pieces.

Let us speculate how the entire process might appear to the student. He or she might see from four to eight individual problems; each of which is at least a bit of a challenge and possibly a significant challenge. Moreover, the methods used might vary from problem to problem in an unsystematic way, and that might be it.

Of course, we are hoping for two more things: (1) that the student sees the interconnections yielding the entire story and (2) that the student learns techniques that are usable in other problems. Is there anything we can do to help? For this particular situation I believe there is (but I think there are many problems that should be rejected because there is little chance, even if the student does all the pieces we place in front of him or her, that he or she will get out of the process anything approaching what we intend). The characterization of Pythagorean triples is nice because the student can learn to appreciate the meaning of the final result by doing simple arithmetic— and one might choose for pedagogical reasons to give away the punch line early by having the student do this sort of arithmetic before launching into the project itself. Insofar as learning transferable techniques is concerned, one can be sure that most students will not realize that they are transferable. This is where some drill exercises can be useful. One does not need to show monotonicity via inequalities for fifty different functions, but after one has learned how for one function while working on the Pythagorean triple problem, then maybe up to a dozen more might be useful; two things are being learned—a technique and the definition of monotonicity.

Let us turn to "problems rather than systematic development." Mathematics departments must be the envy of other departments in institutions of higher learning. Other disciplines strive to give some organized structure to their body of knowledge, but either the structure is extremely elaborate, it doesn't quite fit, or both, and we in mathematics have a subject that by its very nature fits into very nice systematic schemes. Then we say to ourselves that the systematic schemes are boring and do not motivate the students, so let us get away from that and go for problem-solving. I agree with such a sentiment, but a corollary is that then we must make an effort on the organizational side.

Consider, for example, an attempt to introduce students to probability via a game involving repeated flips of a coin. A referee (or one of the two players) repeatedly flips a coin until on three successive flips one of the two sequences THH or TTH occurs; player one wins if it is THH and player two wins if it is TTH. Students are asked which player, if either, has an advantage. It turns out that player two has a significant advantage. From a pedagogical point of view the concern is that even if a student does everything for which one could hope, including being amazed at the conclusion, the student may not have learned any of the mathematical concepts with which he or she is making a tangential acquaintance. Eventually one wants a student to understand stochastic independence in general rather than having just seen it implicitly in examples.

Although I said earlier that group learning can take place whether or not a problem-solving approach is being used, a couple of observations might be made about its use in conjunction with a problem-solving approach. A danger is that the teacher never takes center stage and gives central ideas, such as stochastic independence, their rightful places. An advantage is that since problems have a tendency to be harder than drill exercises, group problem-solving can help avoid a student getting so completely stuck that no mathematical activity takes place.

When I teach using a problem-solving approach, I must remind myself repeatedly that it is much more important that the problems be at the right level than that they be interesting. My excitment should come from seeing the students learn, not from giving them the exciting hint on a problem that was too difficult for them. Students eventually recognize a situation in which they do essentially nothing except follow each of the teacher's hints after a suitable interval of thinking time has elapsed. Of course, posing problems at the appropriate level that are also interesting is better than just making sure that the problems are at the appropriate level.

School of Mathematics, University of Minnesota, Minneapolis, Minnesota 55455

# Obstacles to Change: The Implications of the National Council of Teachers of Mathematics (NCTM) *Standards* for Undergraduate Mathematics

T. CHRISTINE STEVENS

What do the numbers 14, 25, 411, and $\frac{1}{2}$ have in common? Like most good problems, this one has many possible solutions. In this article I will explore one solution—a solution of considerable import for teachers of undergraduate mathematics. The first number, **14**, tells us where American high school students placed in an international comparison of mathematics achievement. When fifteen countries tested their students' knowledge of advanced algebra, the United States placed fourteenth [**6**, p. 22].

## THE TWO CURRICULA

Rankings like this one are obtained by administering to youngsters around the world a test of what I will call the "explicit curriculum"—the mathematical facts and skills that elementary and secondary schools deliberately try to teach. These begin with the basic arithmetic operations of addition, subtraction, multiplication, and division—what students sometimes call "plus, times, take-away, and guzinta." At the elementary level, these operations are applied to whole and fractional numbers, and in secondary school they are extended to algebraic expressions. Also included are such topics as finding perimeters and areas of geometric figures, graphing parabolas, and solving algebra word problems. Facts, skills, and problem-solving abilities like these are what the mathematics teachers in our schools try to teach. Nevertheless, as anyone who has recently taught a freshman mathematics course knows, many students arrive at college without any real understanding of the fundamental concepts in this explicit curriculum.

One reason, in my opinion, is that our schools also unintentionally teach what I will call an "implicit curriculum"—a set of beliefs about what mathematics is and how mathematics problems should be approached. You can discover the essential features of this curriculum simply by asking your C calculus students how they do their homework.

The first tenet of the implicit curriculum is the One-Minute Rule, which states that 85% of all mathematics problems can be solved in sixty seconds or less. From this our students easily deduce that any problem that cannot be solved within five minutes is not worth working on. The reason students believe that mathematics problems can be solved so quickly is that they also subscribe to the Uniqueness Rule, which holds that every problem can be solved in only one way. To find that single correct approach, they compare the problem at hand with examples that have been done in class or in the book. Once they have found the right "template," they simply repeat what was done before, with the numbers changed. The whole process, in their experience, usually takes no more than a minute or two.

Behind this strategy lies our students' belief that mathematics is nothing but a bunch of procedures that they have to memorize. For them, learning more mathematics just means learning more procedures. As one unusually perceptive junior high school student put it, "Math is a set of rules. You have to figure out which one to use. The teacher knows, but she won't tell." Finally, since these procedures are fixed and unchanging, students feel no need to understand *why* they work.

Although students have considerable difficulty with the explicit curriculum, they seem to learn this implicit curriculum very well indeed. The success of this curriculum, in fact, probably helps to account for the second number I listed at the beginning of this article, **25**. Twenty-five is the percentage of students in Calculus I who drop or fail. Another 10% or more get D's and thus emerge from the course with a level of understanding that neither we nor they consider satisfactory [**1**, p. 215].

If the implicit curriculum complicates our lives as teachers, it should also trouble us as researchers. It was not because we liked doing routine computations or because we enjoyed memorizing procedures that we went into mathematics. We became mathematicians because we found mathematics a creative, challenging subject that involves an exciting interaction between intuition and proof and a dynamic interplay between theory and application. For us, mathematics is, above all, a realm where things *make sense*. Students who experience mathematics as a sequence of routine exercises to be solved by standard procedures will never know *our* kind of mathematics. They will never become mathematics majors or mathematics teachers or research mathematicians. Thus the implicit curriculum also helps to explain why so few Ph.D.'s in the mathematical sciences are awarded to American citizens. Last year that number was only **411** [**2**, Table 5].

To my mind, the beauty of the NCTM *Standards* is that they address both the explicit *and* the implicit curriculum in mathematics. They recognize that we will not see much improvement in students' mastery of the explicit curriculum unless we also do something about the implicit curriculum.

## Reasons for Change

Yet schools will not change what they do at the elementary and secondary levels unless we change what *we* do at the undergraduate level. Several factors inextricably link improvement in precollege instruction with changes in collegiate mathematics. In many ways, America's colleges and universities set the agenda for our elementary and secondary schools. Through our entrance requirements, our placement tests, and the expectations we have in our freshman courses, we let the schools know what we consider really important. If we tell them that factoring polynomials, computing square roots, and solving "template" problems are what mathematics is all about, then that is what they will teach.

Colleges and universities also provide the teachers of America with their knowledge of both the explicit and the implicit curriculum in mathematics. Too often, we reinforce *in them* the very misconceptions that we find so frustrating in their students. Many of our tests focus on discrete skills, rather than fundamental concepts, and our tendency to think of teaching as lecturing, and lecturing alone, often allows them to substitute memorization for understanding. We give them, and all our students, very little opportunity to work on problems that require creativity and imagination, rather than mere computational skill. Precollege teachers who have been taught this implicit curriculum will teach it, in turn, to their students. Bluntly put, we get back the kinds of students we graduate.

Yet, if the NCTM *Standards* are actually implemented, we will begin to see very different students in our freshman classes. Prepared by a different explicit curriculum, they would probably be more familiar with discrete mathematics, probability, and statistics and have a deeper understanding of functions and graphs than students do now. They might, on the other hand, be less skilled at verifying complicated trigonometric identities and less willing to treat factoring as an end in itself.

They would also have been exposed to a different implicit curriculum. Accustomed to working in groups on open-ended problems and to thinking for themselves, rather than merely imitating what the teacher does, they would probably be bored by a course that consisted of nothing but lectures and routine homework. They would expect to make frequent use of computers, spreadsheets, and graphing programs, and if we told them to put their calculators away for the semester, they would think us downright silly.

The *Standards*' treatment of probability at the high school level [7, pp. 171–175] illustrates the new kinds of mathematical experiences that these students would bring to college, as well as the mutual dependence of the explicit and the implicit curricula. *All* students, whether they intend to go to college or not, will be expected to wrestle with situations like the following:

> Suppose Anne tells you that under her old method of shooting free throws in basketball, her average was 60%. Using a new

method of shooting, she scored 9 out of her first 10 throws. Should she conclude that the new method really is better than the old method [7, p. 172]?

Mathematicians, who usually approach this problem via the binomial distribution, often object that it is not within the grasp of *all* students. For clearly only a few students will have the understanding of permutations and combinations and of independent and dependent events that is needed to solve the problem in this fashion. This argument reflects, however, a tacit acceptance of the current implicit curriculum. Once the Uniqueness Rule has been rejected, it becomes possible for students who are less comfortable with abstract reasoning to model the situation by rolling icosahedral dice or by using computer-generated random numbers. Thus changes in the implicit curriculum will not only deepen, but also expand, the explicit curriculum.

## Types of Change

I believe that students prepared in this way would be a pleasure to have in our classes. In order to take advantage of their talents, however, college and university mathematics departments will have to change their own explicit and implicit curricula. Sooner or later, these changes will affect virtually all of our teaching activities.

Perhaps the greatest source of change will be technology. Although the appropriate role for calculators and computers in mathematics instruction is still under investigation, it is clear that our courses will have to incorporate these devices in some way—because our students will expect it, because it will make our teaching more effective, and because that's how mathematics is applied in the real world. (When I listen to the passionate arguments that are sometimes made against calculators and computers in the classroom, I cannot help thinking of my own twelfth-grade mathematics teacher, who categorically forbade the use of slide rules and solemnly warned against the dangers of logarithms, more or less on moral grounds.)

This expanded use of technology will both permit and compel a change in *how* we teach. Although I find lectures an invaluable tool for placing problems in context and for delineating the connections between disparate concepts, I also believe that we will not be able to change the implicit curriculum unless we involve students more fully than lectures alone ever can. Students of mathematics should have a chance to hear clear expositions of theorems and to see carefully explained examples, but they must also have an opportunity to make and test conjectures, to form generalizations, and to solve realistic problems. In courses that rely on lectures alone, these latter activities are likely to be given short shrift. By diminishing the time allotted to routine computation, calculators and computers will allow us to devote more attention to these frequently neglected aspects of the curriculum.

These changes in technology and in teaching methods will affect courses of every kind. Obviously we must revise those mathematics courses that are specifically designed for future elementary school teachers. Yet precalculus courses also deserve our attention, because it is in such courses that many freshmen begin, and end, their collegiate mathematics career. As the National Science Foundation's (NSF) initiative underscores, calculus, too, is of great significance, for it is the gatekeeper to further study in science and engineering, as well as mathematics. Even upper-level courses for mathematics majors will require revision, if for no other reason than the fact that such courses are the training ground for secondary school mathematics teachers.

Finally, in the light of new explicit and implicit curricula, we will have to change the ways in which we assess what students know. Placement tests that are confined to algorithmic computation and routine problems will no longer be appropriate, and we will have to devise for our undergraduate courses examinations that measure our success in communicating not only the explicit facts and skills that we want our students to master, but also the implicit habits and attitudes that we want them to adopt.

## PRIORITIES

These changes in teaching, testing, and technology constitute a pretty tall order. No mathematics department, no matter how great its commitment to educational reform, can undertake all of them at once. In fact, no mathematics department *should* undertake all of them at once. Each department must choose, and its choices should reflect its resources (both human and financial), its problems, and its needs. During my recent service as a program officer at the NSF, I observed many schools and colleges and universities as they engaged in the difficult and complex task of changing the instruction they offer their students in mathematics. Drawing upon this experience, I would suggest the following guidelines for mathematics departments that wish to begin the important task of educational reform.

**First, get ready.** Just as in mathematical research, it is important to be familiar with the literature. Obviously, you should read the NCTM *Standards*. You should also read *A challenge of numbers* [4] and *Moving beyond myths* [9], the reports recently issued by the National Research Council's Committee on the Mathematical Sciences in the Year 2000, as well as the latest recommendations from the Mathematical Association of America concerning the mathematical preparation of teachers [5].

Then discuss the *Standards* with teachers and supervisors in the schools from which your students come. Find out what changes they plan to make, and try to analyze the implications of those changes for your undergraduate program. You should also talk with people in business and industry to find out what mathematical skills they want their employees to have. Educational reform is always hard work; but it is easier if you work together with others, and it is almost impossible if you are working against them.

You may also need to make some material preparations for change. As soon as you can, put a personal computer in the office of every faculty member in the mathematics department. Computers will play a large role in the changes provoked by the *Standards*, but many instructors will need to discover the utility of these machines in their own work, before they can explore their applications in the classroom.

**Next, get set.** Decide what problems really *bother* the mathematics department. These days almost all mathematics departments struggle with such problems as inappropriate placement in mathematics courses, high flunk rates in calculus, large enrollments in remedial courses, and a paucity of mathematics majors, but the relative importance of these problems varies from one institution to the next. Whatever changes you undertake will consume some of your department's resources, both human and financial, and the *department* must be convinced that they are worth the cost. Of these two investments—human and material—the former may ultimately prove the more difficult one to make. Having spent two years funding mathematics education projects at the NSF, I realize all too well that money does not grow on trees, but I know that it can be raised in Congress, state legislatures, and local school districts. The time of two or three good people, however, is a precious commodity that is difficult to replace, and it should be invested only in projects in which the mathematics department believes. Do *not*, under any circumstances, let your dean or vice-president tell you what to do.

**Finally, go.** Once you have identified the problems on which you want to work, find some low-risk ways to begin to address them. Undergraduate instructors, like precollege teachers, will find it easier to experiment with new mathematics content and new teaching methods in a supportive environment that minimizes the consequences of failure. They will be more willing to try new ideas if the entire mathematics program does not depend on the success of their efforts. Perhaps the department should organize an honors calculus workshop, introduce a new unit into a precalculus course, or establish a summer program for talented high school students. Do not try to revamp the entire curriculum, explicit and implicit, at once.

Yet even though you start small, you must think big. Try small experiments that could lead to major changes in courses with large enrollments. Changes that are confined to courses serving only a handful of future teachers will be lost in the background noise. Most of our students are freshmen and sophomores, and we must alter the explicit and the implicit curricula that we offer them.

Perhaps the biggest issue of all is minority education. Soon $\frac{1}{2}$ of the school children in the United States will be members of racial minority groups. Although different studies offer different dates for this change, the demographic trend is clear: By the time many of us retire, the majority of America's students will belong to those groups with whom the current mathematics curriculum has been least successful [3, pp. 3-4; 8, pp. 18-21; 10, p. 11]. Our nation will not have all the mathematicians and mathematics teachers (and engineers and chemists and doctors) it needs unless we do something to improve the success of minority students in mathematics. The good news is that several mathematicians, including Uri Treisman and Manuel Berriozabal, have developed programs that stimulate minority students to excel in mathematics. The replication and adaptation of such programs, and the creation of new ones, are of paramount importance.

## The Challenge

The preceding discussion illustrates *one* thing that 14, 25, 411, and $\frac{1}{2}$ have in common: each tells us an important fact about the state of mathematics education today. Yet a good teacher, like a good mathematician, knows that the successful solution of one problem is also an opportunity to pose a more challenging one. Therefore I am going to add **40,000** to my list of numbers—that is the number of people teaching mathematics at American colleges and universities [4, pp. 65, 67, 116]. The challenge is for us and the rest of these 40,000 people to make the choices we need to make about the changes we need to make, in order to make the NCTM *Standards* a reality.

### References

1. Richard D. Anderson and Donald O. Loftsgaarden, *A special calculus survey: preliminary report*, Calculus for a New Century: A Pump, Not a Filter (Lynn Steen, ed.), Mathematical Association of America, MAA Notes no. 8, 1988, pp. 215-216.
2. Edward A. Connors, 1989 *Annual AMS-MAA Survey* (First Report), Notices Amer. Math. Soc. **36** (1989), 1155-1168.
3. Harold L. Hodgkinson, *All one system: demographics of education, kindergarten through graduate school*, Institute for Educational Leadership, Washington, DC, 1985.
4. Bernard L. Madison and Therese A. Hart, *A challenge of numbers: people in the mathematical sciences*, National Research Council, Washington, DC, 1990.
5. Mathematical Association of America, *A call for change: recommendations for the mathematical preparation of teachers of mathematics*, Washington, DC, 1991.
6. Curtis McKnight et al., *The underachieving curriculum: assessing U.S. school mathematics from an international perspective*, Stipes Publishing Company, Champaign, IL, 1987.
7. National Council of Teachers of Mathematics, *Curriculum and evaluation standards for school mathematics*, Reston, VA, 1989.
8. National Research Council, *Everybody counts: a report to the nation on the future of mathematics education*, Washington, DC, 1989.

9. National Research Council, *Moving beyond myths: revitalizing undergraduate mathematics*, National Academy Press, Washington, DC, 1991.

10. The Task Force on Women, Minorities, and the Handicapped in Science and Technology, *Changing America: the new face of science and engineering* (Interim Report), Washington, DC, 1988.

DEPARTMENT OF MATHEMATICS, ST. LOUIS UNIVERSITY, ST. LOUIS, MISSOURI 63103

# The Role of Teachers in Mathematics Education Reform

JOSEPH G. ROSENSTEIN

ABSTRACT. In discussions of the state of mathematics education today, teachers are often viewed as being part of the problem. More attention should be focused on how teachers are and can be an important part of the solution. Drawing on programs directed by the presenter, this paper will discuss some of the roles for teachers in mathematics education reform.

The above abstract, prepared several months before this paper, appears to fit both as an introduction to and as a summary of this paper.

## I

First of all, when we mathematicians think about mathematics education, our view of high school mathematics is typically that the high school teacher is part of the problem.

What else should we think when the students we see as first-year undergraduates are lacking in the mathematical skills we think they should have. Why can't our students add fractions or remember the rules for exponentiation? Why can't they identify, let alone sketch, the graphs of basic functions? Why do they have such trouble with word problems, and why don't they recognize that their answers to word problems are often simply nonsense?

"Obviously," many of our colleagues conclude, "it must be the fault of their high school teachers. They had these students in their classes just last year. Why didn't they teach them these topics?"

The strange thing is that when senior-level high school teachers get together they ask the same questions—but they address them to teachers of intermediate algebra... and we can guess what happens when intermediate algebra teachers get together!

Everyone agrees that many students do not achieve mastery in mathematics, but the causes are complex; we could easily enumerate a half-dozen

---

This paper is the result of a presentation given at the special session of Mathematicians in Education Reform (MER) Network, held during the Joint Mathematics Meeting in Louisville, Kentucky on January 18, 1990.

differences between the 1990s and previous decades which would help to explain the phenomenon. Focusing simply on the high school teachers as the cause of the problem is an example of the *post hoc ergo propter hoc* fallacy we have all learned to avoid.

There are, of course, many high school teachers of mathematics whose training is not adequate, often through no fault of their own. They may have been pressed into service due to a shortage of mathematics teachers in their states—or due to a surplus of social studies teachers in their school systems. They may be graduates of teacher training programs which emphasized pedagogy and all but ignored mathematics. They may have returned to teaching—or to mathematics—after many years elsewhere. Or they may simply have not learned mathematics well.

But there are also many expert high school mathematics teachers. While I cannot make any statistical claims about their numbers, as a result of my activities during the past five years, I know that there are many. I have worked with them on a number of projects and programs. Their enthusiasm for mathematics matches ours, and their desire and ability to communicate that enthusiasm usually surpasses ours.

Unfortunately, these teachers are not in the public eye; the reports on education focus on the shortcomings of mathematics education, not on its strengths.

Expert teachers are an important, often untapped resource in the reform of mathematics education. In this article, I will describe several models of how their expertise can be used, as well as how the academic and research mathematics communities can facilitate that process.

## II

Not only is society unaware of the existence of expert mathematics teachers, we mathematicians do not recognize their existence either.

The split in the mathematical community between high school and college-level teachers has caused a great deal of harm to both. There may never have been a time when college teachers really saw high school teachers as colleagues,but for the last few academic generations, there has been no awareness among academic mathematicians that the two groups share a common background and a common agenda.

How often does it happen that high school and college mathematics teachers sit down together and discuss, as colleagues, their common concerns about mathematics education?

For most of us, the only encounters we have with high school mathematics teachers is at parents' night at the high school, and our typical reaction is that we would do a better job of teaching our kids.

We are often disappointed when we find that high school teachers are not as knowledgeable or skilled in mathematics as we are. We forget that we need them to be teachers, not research mathematicians, and that their training is not the same as ours. We also assume, incorrectly, that because we know mathematics better, we can teach it better.

Consequently, the proposals that come from the college community are often prescriptive. "They don't really know about mathematics and how to teach it. We'll show them how to do it right!" is a common attitude among our colleagues. Another is "Let's get better mathematics students to become high school teachers!" Such proposals don't go very far in building a collegial atmosphere. That's not to say that we shouldn't upgrade the skills of teachers and that we shouldn't encourage more good mathematics students to become teachers, but if that's all we say, then we project a perspective that is bound to be counter-productive.

We have to start, instead, with the assumption that many high school teachers of mathematics share our enthusiasm for mathematics and share our agenda for reform and improvement in mathematics education.

We have to start with the assumption that high school teachers often know as much, or even more, about the problems than do practicing mathematicians.

We also have to accept the reality that changing the teaching of mathematics in the schools can only be effected by teachers in the schools.

We must recognize that high school teachers must have important roles in mathematics education reform and use our resources and our stature to help them bring about that reform.

We must also encourage and support their efforts and work toward real partnerships in mathematics eduction reform, where the pronouns "we" and "they" are no longer used in a self-congratulating or pejorative way.

## III

Three examples of such partnerships will highlight three important roles that teachers can have in mathematics education reform—in training inexperienced teachers, in disseminating new approaches, and in cooperative efforts in improving instruction.

These roles are highlighted, respectively, in the Institute for New Mathematics Teachers, the Leadership Program in Discrete Mathematics, and the Precalculus Project, three programs I direct at Rutgers University. Brief descriptions of these programs are provided here; further information can of course be obtained from me directly.

These programs are sponsored by the Rutgers University Center for Mathematics, Science, and Computer Education. The Center was founded in 1985 to work with schools in the areas of mathematics, science, and computer education. Directed by Gerald A. Goldin, the Center sponsors research, curriculum development, summer institutes, partnerships with school districts, conferences, lecture series, and in-service educational programs. The Center has been funded since its inception by Rutgers University and the New Jersey State Department of Higher Education.

**One very successful program involves using expert mathematics teachers to help train new mathematics teachers.** The Institute for New Mathematics Teachers (and a parallel institute for new science teachers) is a week-long residential institute which takes place each August, with follow-up sessions during the school year. The program has been funded for the last three years by the New Jersey Department of Higher Education using Title-II funds, now called Eisenhower funds.

Participants bring with them their teaching assignments for September and review with the staff, in small homogeneous groups, the textbooks and curriculum guidelines they will be using for their classes. The staff provides the new teachers with suggestions for how they can supplement the material and for how they can motivate their students to learn. The title of the institute is Teaching Excellent Mathematics, and that is a goal that we hope will ultimately be achieved by the ninety new mathematics teachers who have participated in the first three years of the program.

The staff consists of high school and middle school teachers and supervisors of science and mathematics who are selected because of their own reputations for excellence in teaching.

Here are a few reasons why excellent practicing teachers are particularly appropriate for this type of role:

- They have collected and developed materials over the years which they have used in their courses, and they are eager to share these materials with the new teachers.
- They have been able to show their own students that "mathematics is not a spectator sport," and they have been experimenting with current ideas and techniques such as "discovery learning," "problem-solving," "hands-on manipulatives," and "cooperative learning" to enhance student involvement in the learning process.
- They are enthusiastic about mathematics and caring about their students and want to communicate their enthusiasm to their students.
- They serve as excellent role models in instruction; they are able to involve the participants in the program in the same ways that they would like the participants in the program to involve their own students.
- They are mentors who can explain their own commitments as teachers of mathematics and can provide the new teachers with assistance and support in becoming expert teachers.
- They are colleagues who have the same students and administrators to deal with as the new teachers. They are dealing every day with the same kinds of situations that the new teachers will be facing, so the new teachers can easily identify with them.

We have been truly fortunate that each staff member selected for our program has indeed been an excellent selection, and when you think in terms of the varied roles that the staff play in the program, you can see how important this has been for the success of the program.

I'm certain that there are many other teachers who would also have been excellent selections. I have learned that **there are many teachers who are really** *under-utilized*, **that the structures of schools and school systems do not provide these expert teachers with opportunities to share their expertise.**

**We need to identify these teachers and help create such opportunities for them.** We mathematicians and our institutions are in a position to make such opportunities happen; because of our reputations, we can attract the participation of expert teachers as staff, of learning teachers as participants, and of state and federal agencies as financial supporters.

Let me expand on this point. Suppose that you decide to organize a series of workshops for high school teachers, and that you set about to identify and select the best teachers in your area as staff for the program. These teachers would want to participate because their selection by a college or university to play such roles is itself welcome and unexpected recognition of their efforts and their expertise. Other teachers would want to participate because your institution has a reputation in your state, and both the participants and their supervisors would recognize the potential value of the program. Your institution would value the program as an opportunity for both public service and public relations, and the funding agencies will be interested because you are providing an opportunity that no one else can.

A brief discussion of another example: As part of our Precalculus Project, about which I will say more shortly, we offer a summer institute for precalculus teachers. We find that high school teachers often spend many years teaching lower-level courses before they have an opportunity to teach precalculus—often when a precalculus teacher retires. By that time, they need a real conceptual review of the material and an opportunity to explore how the apparently disparate pieces of precalculus fit together. The summer institute provides such an overview of precalculus to twenty-five teachers each year, and, to a large extent, it is staffed by expert high school teachers.

In these examples, teachers teach other teachers about traditional components of the high school curriculum, but **the principle of teachers teaching teachers is applicable also when new areas and focuses of mathematics are discussed.**

Let us assume that the community has decided that a new topic—perhaps probability and statistics, or discrete mathematics, or mathematical modeling, or linear algebra—or a new technology should receive increased emphasis in the schools. How should we carry out that decision? Simple arithmetic shows that even if each of us were to conduct a program for teachers, the number of teachers we could reach would be very few indeed.

In order to involve many schools in a new direction, teachers have to be enlisted to instruct additional teachers. That means that we should not focus exclusively on upgrading the skills of individual teachers—as did National Science Foundation (NSF) funded projects in the 1960s—but to work with those teachers who can most effectively carry the message to other teachers.

This is one focus of another program that we offer at Rutgers University. Early in 1989, the NSF awarded a grant to create a Science and Technology Center in Discrete Mathematics and Theoretical Computer Science (DIMACS); located at Rutgers University, DIMACS is a consortium of Rutgers and Princeton Universities, AT&T Bell Laboratories, and Bell Communications Research. As part of its educational program, DIMACS provides support for an institute in discrete mathematics for high school teachers. The first program, entitled Networks and Algorithms, and funded by DIMACS, took place in the summer of 1989; an expanded version of the pilot program, called the Leadership Program in Discrete Mathematics, was offered in the summer of 1990 with financial support from the NSF.

The purpose of the Leadership Program in Discrete Mathematics is to create a leadership cadre of high school teachers. We expect program participants to become knowledgeable about concepts of discrete mathematics and their applications. We also expect them to develop materials and activities for incorporating these concepts into their classes and to introduce their colleagues to these materials and activities.

We cannot, of course, assume that participants will come to the institute knowledgeable about discrete mathematics, but we can ensure, through the recruitment and selection process, that the participants are teachers who will be able to play leadership roles in disseminating information to colleagues both in their schools and in other schools.

During the institute, participants study the subject matter and applications of discrete mathematics. They then work together to create materials and lessons which they will use in their own classes. At follow-up sessions they discuss their activities with each other and report to the group on their progress. Once they have successfully incorporated institute topics in their own classes, they find suitable forums, within their districts and beyond, for disseminating the topics to other teachers.

Although the pilot program has not yet been completed, it is clear that it has been successful. Essentially all of the participants have indeed incorporated institute materials into their classes; many have been preparing curriculum materials based on what they have done; and many have given, or are planning to give workshops for other teachers.

Those participants who are most successful will be offered broader responsibilities, as "lead teachers," at subsequent institutes. Drawing on their own experiences in teaching discrete mathematics and in incorporating new topics into their curricula, the lead teachers will be able to provide assistance and advice and to serve as role models to the teachers in the program. Ultimately, we expect the lead teachers to lead workshops for teachers throughout the region on how to introduce discrete mathematics into the high school curriculum. This model and the theme of "teachers teaching teachers" was developed for the Woodrow Wilson Institutes that take place each summer at Princeton University under the sponsorship of the Woodrow Wilson National Fellowship Foundation.

**The theme of the Institutes for New Mathematics Teachers is to use experienced teachers to teach new teachers, and the theme of the Leadership Program in Discrete Mathematics is to use lead teachers to disseminate new approaches. A third program, the Precalculus Project, has a different theme altogether—that of teachers working together to improve instruction.** The Precalculus Project grew out of one of the first activities sponsored by the Rutgers University Center for Mathematics, Science, and Computer Education.

Consistent with its mission, the Center's first event was a conference in March 1985 attended by mathematics, science, and computer educators from around New Jersey.

Those attending the conference were asked in which areas they particularly sought our assistance and involvement, and the mathematics teachers and supervisors responded that one area which needed particular attention was precalculus. Indeed, when fifteen high school teachers and mathematics chairs gathered subsequently, they discovered that each of their schools had a precalculus course, but that no two of their courses were the same. The group formed the Precalculus Committee and set about determining what students needed in order to be properly prepared for calculus.

After a year of monthly meetings, the Precalculus Committee produced a report called *The Core Precalculus Course*, which described in some detail what students needed to understand and be able to do in order to succeed in the first two semesters of calculus. Recognizing that different schools and different types of students have different needs, the committee did not attempt to prescribe a standardized precalculus course; rather, the committee tried to describe what the "core" of every precalculus course should be. In some situations, that core could be completed in one semester and supplemented by additional material; whereas, in others, a full year might not even suffice. The report was disseminated throughout New Jersey and has been used by many districts in reviewing and revising their mathematics curricula.

Since the focus of the article is not on precalculus but on the role of high school teachers, I will return to the main topic. As the committee proceeded with its deliberations, I found that the teachers were attending our late afternoon sessions regularly, sometimes braving the New Jersey elements to do so. I asked myself why were they so devoted; why were they coming to our meetings after a full day of teaching? After a while, I realized that they weren't coming just because it was important to complete the report. They were coming because they found that the meetings were themselves valuable.

Discussions of mathematics and the teaching of mathematics are unfortunately rare in the schools. Mathematicians who teach in colleges and universities take for granted the existence of colleagues and the opportunity for collegiality. If we have a mathematical concern, we simply walk down the hall and bounce our ideas off our colleagues; or we have our discussion over lunch; or we call a meeting.

A high school teacher often does not have colleagues. He or she may be the only precalculus teacher in the school, or, equivalently, the only precalculus teacher who cares about mathematics or about education. Even if there are other teachers, their schedules may preclude any conversation other than brief exchanges between classes. Lunch is a time for cafeteria duty, not for professional talk, and departmental meetings are usually devoted to administrative, rather than educational concerns.

As a result, many high school teachers do not have opportunities to engage in the kind of conversations that the Precalculus Committee offered—conversations that deal with mathematics and how one can provide a setting where students will be motivated to think mathematically.

I found that our discussions of the core precalculus course were providing exactly this opportunity, so we have followed this up with other opportunities. For example, we have a test item bank committee, which has been working for several years on creating a collection of precalculus questions based on the core precalculus course recommendations. The committee distributed the first version of the "test item bank" two years ago and each year since has worked on improving and extending various sections of the original test item bank. The meetings of the committee, while focused on developing a "product," provide the participants with interesting discussions on precalculus topics. At a recent meeting, the following questions were discussed: What kinds of inequalities should we expect students to be able to solve? What techniques do we expect them to use? If we emphasize the interrelationship between the geometric and algebraic perspectives, should we spend more time on solving inequalities graphically? Where else can teachers discuss such questions!

Each year we sponsor a day-long conference entitled Good Ideas in Teaching Precalculus which is attended by over 200 New Jersey high school and college teachers of precalculus. At each of five sessions, participants can choose to attend one of about eight programs, many of them presentations by high school teachers. At each session, we also have "idea exchanges"—opportunities for teachers to engage each other in discussion on a particular topic—for example, introducing limits in precalculus. In both the presentations and the idea exchanges, teachers have an opportunity to discuss mathematical concerns with their colleagues.

This spring we are enlarging our program to create several new workshop groups. One will deal with sequencing precalculus topics, another with motivating precalculus students, and a third with implementing in the precalculus course the *Curriculum and Evaluation Standards* which was recently published by the National Council of Teachers of Mathematics (NCTM). Each workshop group will meet on four weekday afternoons during the spring and will formulate an objective to be achieved in these four meetings. Two types of benefits will be achieved through these meetings. The participants will have an opportunity to discuss mathematics and the teaching of mathematics with their colleagues, and the community will benefit because what each group produces will be disseminated to other teachers in the state.

This year we are also planning to introduce several more intensive workshops, involving five full days of activity. For example, at any given time there are many districts that are revising their precalculus curricula, or that are considering districtwide implementation of the NCTM *Standards*, or that are considering the introduction of new material in the precalculus course. Why not bring people from these districts together so that they can benefit from joint discussions about curriculum revision (or one of the other topics)? Why not have this group draw up a handbook of issues that need to be addressed in revising a precalculus curriculum?

Each of these activities can be considered as an opportunity provided, under university auspices, for high school personnel to have professional discussions which will be useful for themselves and for the broader community.

The Precalculus Project has grown from a small committee to a mailing list of over 1000 New Jersey precalculus teachers who have participated in our programs or requested our materials. It is now an independently funded program, supported by Title II (Eisenhower) funds provided by the New Jersey State Department of Higher Education. Many high school teachers are actively interested in improving mathematics education and in working and networking with their colleagues. The Precalculus Project provides them many opportunities for doing so.

## IV

Drawing on programs I direct at Rutgers University, I have described three roles for high school teachers in mathematics education reform—in training inexperienced teachers, in disseminating new approaches, and in cooperative efforts in improving instruction.

I hope that I have argued convincingly that there are dedicated and enthusiastic teachers who are interested in playing such roles, and that providing opportunities for teachers to do so is a worthwhile endeavor.

It is an endeavor to which I have devoted most of my time and efforts in the last few years. After writing a number of articles in model theory and recursive function theory and a book on *Linear Orderings*, I served as Director of the Undergraduate Program in Mathematics at Rutgers University, New Brunswick in the early 1980s. By instituting a placement test, we found out what mathematics incoming freshmen knew (or didn't know). As a result, I became involved in a number of university and statewide efforts to formulate expectations for college-bound students. Naturally, the next step was to consider how those expectations could be realized. At the time, the appropriate assumption seemed to me that high school teachers would be allies, not adversaries, in the process. As I became involved in various activities, I repeatedly found evidence that my assumption was indeed appropriate. So it is not surprising that I am now organizing opportunities for teachers to bring about change in their instruction and in their schools.

Organizing such opportunities is not a simple task and may be much more than most readers of this article are willing to undertake. Nevertheless, there are an increasing number of mathematicians who are involved in such organizational efforts and who would welcome the interest and participation of their colleagues. In any case, there are two tasks with which I would like to charge all of my readers.

The first is to counter the negative remarks that are continually directed at high school teachers; don't let your colleagues get away with gratuitous potshots at the schools! Don't let them generalize from personal anecdotes to a general indictment of all high schools and all teachers. Such generalizations are academically dishonest and clearly counterproductive; like other prejudices, they should be opposed vigorously. Positive change can only come about if we and our colleagues recognize that high school and college teachers of mathematics have similar problems and similar goals.

The second is to find ways to provide support and encouragement to teachers. Visit your local schools and sit in on some classes; you may even be invited to try your hand at conducting a high school class. Be supportive—even though you may not feel supportive. Acknowledge that the teacher's job is not an easy one and offer your encouragement. Recognize that the curriculum can be improved and encourage the school to explore alternatives, but be aware that you may not have the background to be sure that your solutions are the right ones. Be particularly cautious about recommending broad changes if your main goal is to improve your own child's education; what you think is best for your child may not in fact advance the common good (and may not even be best for your child). Show that you care and appreciate the teachers' efforts, that you are willing to serve as a resource and a colleague, and that you are willing to put your time and thought into a common search for solutions.

DEPARTMENT OF MATHEMATICS, RUTGERS UNIVERSITY, NEW BRUNSWICK, NEW JERSEY 08903

# Teaching to Love Wisdom

CHRISTOPHER COTTER AND IGOR SZCZYRBA

SUMMARY. This article attempts to share our ideas on the reform of Mathematics Education. We describe the structure of a new Ph.D. program in Educational Mathematics that implements our views.

*Doctoribus atque poetis omnia licent*

## MOTIVATION FOR OUR NEW APPROACH IN EDUCATIONAL MATHEMATICS

It would be very interesting to check how many readers, who hold Ph.D. degrees, recognize that the title of this article refers directly to that degree. Does the majority of academia still associate the term *University Doctor* with its historical meaning? Let us remind those without a handy dictionary: *doctor* means 'teacher'; *philosophy* means 'love of wisdom'. By starting with this linguistic question, we wish to emphasize that within our community it is not clear what role Doctor's of Philosophy in Mathematics should play in the current and future educational system.

Should we, as it is advocated by some mathematicians, for instance at the Annual Meeting in Louisville in January 1990, *constrain* our educational activities within the traditional model which assumes that doctor's of mathematics mainly do research and pass their knowledge to a relatively small group of gifted pupils who in turn become a new generation of Ph.D.'s? This elitist solution has been working for several hundred years, and we agree that mathematicians should not be afraid to enhance this elitist approach. There is nothing wrong in stressing that mathematical talent is a gift of mind as precious as, for example, musical talent, and as such, it should be cultivated in a special way. We all know that the best musicians take an active role in guiding young musical talents at as early an age as possible. Shouldn't mathematically gifted children be taught by the best mathematicians?

Some aspects of this question were discussed by Dr. H. Clemens in his article *The Scarlet E* (UME Trends, January 1990). His idea to give university credit for teaching interested school children is very sound and has a tradition. For instance, Felix Klein's efforts to bring a higher prospective of mathematics into the teaching of school children—and who can better do this than a mathematician himself. The involvement of university mathematicians in school education has even been institutionalized in some countries in the form of special classes or schools.

But will our involvement with small groups of eager students be sufficient to solve the crisis in mathematical education in the United States? We do not think so given the likely number of participants. In fact, Dr. Clemens' own lament—that the American research community views mathematical education in the light that "too many mathematicians **do** in fact use math education, ..., as a way to escape doing mathematics"—is somewhat ironic. The real problem is that the mathematics research community has for a considerable time avoided active involvement in mathematics education at lower levels and so, given up any significant control over elementary and secondary math education.

In consequence, similarly as it happens in a biological environment, other species evolved to dominate the existing niche—mainly general educators and science educators. They were trying to do what they believed was the best. In particular, in the preparation of teachers they emphasized their field, i.e., methodology or very specific applications, over the mathematical content, and it was not a matter of good or bad will. They could not do anything else since they were rarely exposed to mathematics in a proper way. Consequently, most school teachers do not understand how to enjoy and deal with mathematics the way we do.

Of course, one can say that gifted students survive anyway; that they are still entering colleges enabling us to preserve the old tradition of masters and pupils. But many are lost to mathematics forever who could have become excellent researchers and teachers.

Mathematicians, as a professional group, have failed to find a solution that:

1. would 'produce' sufficiently many well-trained mathematicians at various levels; and,
2. would continue to update the knowledge of currently active teachers.

We believe that it is crucial for the future that the initiative *in solving these two problems* be taken into the hands of mathematicians. That means, in particular, that the new strategies in educational mathematics should be created by mathematicians. Moreover, mathematicians should carefully monitor all ongoing programs in mathematical education. We are aware that many mathematicians share this point of view and have been developing valuable projects for many years. The reader should also admit though that what has been achieved in the last ten years to upgrade mathematics education on a global scale isnot too impressive. For instance, the vast majority of the articles edited by Dr. L. A. Steen in *Mathematics Tomorrow* (Springer 1981), could be reprinted now, without any changes, in any journal that discusses current educational reform.

It is even more frustrating to admit that the accomplishment is so little, given that (according to the official statistics for 1988 of the Department of Education) the United States spends per capita on education the second largest amount in the world after Switzerland. In fact, one would surmise from current surveys describing competency of American students that much of these public funds were misspent. (And we should not forget about additional private funds that were donated 'to improve' the American educational system.) Now, it seems that society as well as those administrating it are willing to spend even more money especially for mathematics education. Only mathematicians can insure that these huge amounts of money are being spent in a more advantageous way.

This point of view is shared also by many educators. For instance, Dr. L. Shulman from Stanford University pointed out in a presentation at our university in 1989 that "the school curriculum is an Arts and Sciences curriculum and therefore the education of teachers should be done in these Colleges." Moreover, answering a question, he made it very clear that in his opinion the role of colleges of education in teachers' preparation needs to be significantly diminished.

Thus, in the near future, we shall need more mathematicians who will be deeply involved in the education of mathematics teachers and in the administration of educational reform at all levels, i.e., at universities, four- and two-year colleges, and at school districts. Individuals who are capable of doing all this properly must have a broad knowledge in mathematics and, at the same time, a good understanding of how people learn mathematics. The publications *Everybody Counts* or the new NCTM *Standards* themselves will change nothing without the whole academic community, including research mathematicians involved in the education of such individuals.

It is the common view that the only acceptable way to 'create' the individuals we will need is through degree programs based on mathematics content. Depending on the audience to be targeted, the traditional mathematics degree program could be modified, for instance, by incorporating appropriate knowledge from cognitive sciences. By incorporating we mean something more than just adding to existing programs some extra courses taught by educators. We believe most traditionally required educational courses must be dropped unless they are designed anew. For other audiences, we felt a more significantly altered program was necessary.

In this spirit we created a new Ph.D. program in **Educational Mathematics**. Before we present a more detailed philosophy and structure of this program let us explain the origin of our motto and some deeper reasons for which we decided to use it.

The motto is a quote of a character of the Russian writer N. A. Nekrasov and might be translated in the following way: "Doctors as well as poets are allowed [to do] anything." It is a paraphrase of Horace's: "*Pictoribus atque poetis quidlibet audendi semper fuit aequa potestas*" that represents the views of his supposed opponents: "Painters as well as poets since olden times have had the right to challenge anything."

We use this quote to stress that in the American academic system the doctorate is the preeminent degree and as such provides its holders with great possibilities in influencing any reform. Contrary to the German academic system (that has also been adopted by many European countries, e.g., by the Soviet Union), a Ph.D. graduate in the United States is allowed to teach a broad spectrum of courses and to make important administrative changes from the outset of his career. This makes him a key player in any educational reform. (In the German system, individuals without the higher degree of a *Habilitation* have rather limited influence in academic affairs.)

The choice of our degree and its name is, in fact, a result of our beliefs and ever present administrative constraints. Namely, in the fall of 1988 our department was asked to develop a Ph.D. program proposal to replace an existing D.A. in mathematics that had been effectively moribund for some time. Our natural inclination was to establish a traditional Ph.D. program in mathematics, but due to peculiarities of Colorado Law, the University was only eligible to offer education related Ph.D. programs. (Thus our desire to have a Ph.D. program inexorably led us deeper into the field of mathematics education reform.)

We could have simply adopted the conventional Ph.D. program in Mathematics Education, but we are interested in mathematics—and we still **do** mathematics. Moreover, the syntactic structure of mathematics education, we felt, represented a much deeper, dangerous idea, implicitly accepted by many in the field of education, that subject matter plays the role of modifier of the educational method—pedagogy! We felt very strongly that to the degree it is possible this view needs to be reversed; teaching strategies, which are metaphors for that being taught, need to be derived from the material to be presented. Thus the program was conceived and titled Ph.D. in Educational Mathematics.

In preparing anyone to actively integrate knowledge of their subject into teaching, we accept certain facts of life. That we teach courses as they were taught to us is now something of a cliché in education reform, but it is the case and, for that matter, a good one. Few people can be realistically expected to make a major time-consuming effort in the direction of changing their teaching habits—reeducation for everyone is highly inefficient! Thus in our Ph.D. program we decided to make the priority teaching mathematics while at the same time maintaining a dual line of discussion about its educational value at all levels. This is intended to be a very self-conscious attempt at rearranging the mathematical concepts in our students' minds, so as teachers themselves, the natural 'learned' method of delivery of the concepts will be as closely aligned as possible to what research in cognitive processes and our own experiences can tell us will be successful.

We would like to first state a drawback of this strategy, namely that less material will be covered in the courses. In our program though, we expect our doctorates to be primarily teachers, in particular, teachers of teachers. In this way a cascade effect of significant proportions is anticipated which can have a significant impact on the education of school children.

On the other hand, as far as our final product is considered, we did not want to restrict ourselves just to 'producing' teachers of undergraduate mathematics for two- or four-year colleges. Although we are aware that this is an essential link in the mathematical education chain, we do not think that even an educationally oriented Ph.D. program in Mathematics can be *ex definitione* so narrowly specialized. We believe that some, at least basic, understanding of what mathematical research is all about is a necessary condition for becoming a good math educator. At the same time the 'research' knowledge of our graduates might also be utilized beyond direct teaching, for instance, in administering the reform of mathematics education.

## The Program

The program's structure is the following:

- mathematical core (minimum 28 semester hours),
- educational core (minimum 12 hours),
- electives ( up to 11 hours),
- dissertation.

In addition, students must meet general requirements of our graduate school, such as proficiency in a foreign language.

In the first year, students are expected to complete the majority of the mathematical core. In particular, in this first semester of our program we are offering the following courses: Real Analysis taught from Royden's book, Algebraic Number Theory taught from Marcus' book, Commutative Algebra taught from Atiyah and MacDonald, and Logic taught from the book of Chang and Keisler. In the second semester students will complete Complex Analysis, Differential Geometry, and a course on Topics in Modern Mathematics. These courses will be at a similar level as those above. The second semester also includes two seminars. One will be a research seminar that is linked to one of the first-semester courses. This year Computational Number Theory will be the topic. It is our goal that every student have an experience that is computationally intensive in an essential way to fully introduce them to mathematical applications of computers. Included in this year's seminar will be recently developed methods of factoring integers, primality testing and other computational methods in number theory.

The second seminar, K-12 Cognitive Processes, is a part of the educational core. In this seminar, we do not only intend to present our students with general knowledge and methods of educational research in cognitive sciences. The seminar will also be attended by several of our faculty, and we plan it to be a forum for discussion (we hope critical and creative) about what, where, and how specific mathematical content can and should be presented to grade-school children. Special attention at this seminar will also be given to constructivism and the current research in this field. In particular, we will discuss how the notions, introduced by Piaget in his theory of mathematical education, can be generalized to other aspects in education and the sociology of education.

As we have already mentioned above, the basic tenet of our approach is that mathematical content is fundamental to understanding educational methods. Thus there will be a direct link between the material presented in the content lectures and questions discussed at the educational seminars. In fact, we foresee it as a two-way process, in the sense that during content lectures, whenever it is proper, educational aspects will also be emphasized.

In the second year the students will complete their math core by selecting a few courses from those offered. These might include Representation Theory, Difference Equations and Chaos, Measure Theory and Distributions, or Symmetry in Mathematics. In the last course we intend to cross over many fields of mathematics and to discuss various applications of the notion of symmetry. It is of particular pleasure to note in the first months that the program has been running, that the students have shown a definite desire for solid mathematical content in their courses and that they appear to be capable of handling it.

In addition to the math course work, the students will complete the educational core in the second year. Namely, they will attend the seminar on Post-secondary Cognitive Processes and a seminar on the philosophy of educational mathematics. Moreover, students will be required to complete at least six hours of educationally-oriented courses. For instance, in the three-hour class Mathematics Course Development they will be asked to develop, from scratch, a specific course or portions of specific courses designed for a given audience. The project will include the basic ideas of the correlations between the course, and other courses, and appropriate methods of presentation. More importantly, the class will carefully address the question of anticipating student preconceptions (right or wrong). The goal is to carry this out in a pragmatic manner and not imply that the teacher's understanding of the maturation process can be directly transferred to the student. After reviewing our Ph.D. proposal, Dr. A. Schoenfeld informed us that he teaches a similar course at Berkeley; and, in his opinion, it will be an excellent piece of our program.

Another nontraditional course, Mathematical Structures and Education, will analyze different ways in which basic mathematical structures, already known or new to the students, can be incorporated into various school curricula. This course, as all courses in the educational category, will stress cognition of mathematics. Professor Schoenfeld has pledged his help to guide us in this field. We are in a process of 'soliciting' a larger group of advisors for our program.

The electives can be chosen either from the two categories above or from a pool of graduate courses in statistics, the sciences, computer science, and education. The choice of electives is intended to be closely related to the dissertation topic. Namely, we believe that the electives should either deepen the student's knowledge in the field of his dissertation or help him as a research tool. We assume that students will complete all the course work within four or five semesters. We hope that as the program grows future students will have more flexibility in sequencing their courses.

The types of dissertations that are possible include:

1. pure mathematics with a chapter on related educational problems,
2. mathematical models related to education or administration of education, and
3. cognitive problems in mathematics.

## THE TOUGH BEGINNING

This program was officially approved during the Fall Semester of 1989 after more than a year of preparation. Due to this late date, we were not able to formally advertise its existence. Yet we received more than twenty-five inquiries about our program based on word of mouth and flyers we produced on a PC. We accepted seven applicants into the program. Again though, we were delayed in offering these admissions until May 1990 due to clerical errors in the Graduate School, and then even further delays occurred due to the lack of final authorization to offer the teaching assistantships we were promised. Thus, by the time the offers were sent out several of the students had taken other positions, and we enrolled three full-time students in August 1990. We are continuing to receive many inquires from very well qualified individuals for the future, and it is already clear that future years' enrollment will be higher.

## FINAL REMARKS

We would like to address another fact of life that we have come to 'accept' which is the lack of passion for mathematics among our future school teachers. At our university the majority of undergraduate mathematics majors are prospective secondary school teachers, and one learns very quickly that they do not love mathematics. There certainly is no a priori reason why this is their fault. In fact, it seems a very logical development given the nature of their mathematical experiences with an educational system that accepts the classification of mathematics as a 'hard' (read unpleasant) subject. That they then go on to become teachers produces a highly reinforcing cycle. It is our hypothesis that this is the very root cause of the need for mathematics education reform.

However, our contacts with teachers who are in our master's degree program suggest that it is the lack of a proper understanding of the content which makes it difficult for them to enjoy mathematics. We have observed a significant change in their feelings toward mathematics after they have understood a general concept and were able to realize why and for what purpose they teach it at their schools. For instance, one of our master's graduates wrote that initially she was not especially motivated to get her master's in mathematics, but after she began a deeper study of subjects, she became "proud of mathematics as a topic and [of] teaching mathematics."

In his essay on number theory published in *Mathematics Today, Twelve Informal Essays* (Springer 1978), Dr. I. Richards wrote that Leonhard Euler was so persistent in his work since "great mathematicians are professionals, and professionals love their craft." But maybe, we all love mathematics just because we understand it, and this understanding is the key for *Teaching to Love Wisdom*. We hope that by providing our students with knowledge it is possible 'to contaminate' them with our passion toward mathematics and that they will transfer our love of mathematics' wisdom to their students.

DEPARTMENT OF MATHEMATICS AND APPLIED STATISTICS, UNIVERSITY OF NORTHERN COLORADO, GREELEY, COLORADO 80639

# Teacher Networking: A Corollary of Junior Mathematics Prognostic Testing

FRANK L. GILFEATHER AND NANCY A. GONZALES

## INTRODUCTION

Poor mathematics performance at the secondary and college levels is a critical problem facing our nation's educational institutions today. For years, many universities and colleges throughout the country have sought to remedy the students' deficits in mathematics skills by providing courses of a remedial nature. Departments of mathematics have now begun to realize that these programs have, for the most part, yielded unsatisfactory results. The fact is that remediation through a college semester course (or two) is not an adequate substitute for the conceptual development that takes place within the high school mathematics curriculum. University and college math faculty generally agree that the student is best prepared if he or she acquires the fundamental skills of mathematics *before* high school graduation and not as part of the college program of studies.

Postsecondary institutions of New Mexico have witnessed a woefully high demand for remediation in mathematics. Recognizing the seriousness of the problem, the University of New Mexico (UNM), Department of Mathematics and Statistics, began addressing the issue by initiating a high school junior testing project based on the Early College Mathematics Placement Testing Project of Ohio State University. These testing programs inform high school juniors where they would start relative to math placement, if they were to be enrolling that fall at universities and colleges. National experience shows that as a result of these tests, students enroll in more math classes as seniors in high school.

Mathematics educators from throughout New Mexico felt that the students in New Mexico could benefit from an early and realistic analysis of their mathematics skills—relative to the college mathematics curriculum. Further, the results of such early placement tests motivate students (especially those interested in postsecondary education) to take the appropriate mathematics course during their senior year and, indeed, take these senior level courses more seriously. The testing and related programs begun in 1988 are administered by the Department of Mathematics and Statistics at UNM. The program and tests are free to the schools and students. For a detailed description of efforts to create and pilot early placement testing programs, see [1].

Aside from the primary benefit of the tests, namely, increasing the preparation and success of high school students in mathematics, the New Mexico junior testing program has been designed in such a way so as to bring about an awareness of the teaching of mathematics as a *profession*. Such a view requires specialized knowledge of the teaching and learning of mathematics, which evolves not only through academic preparation but through years of classroom experience. By promoting a sense of professionalism, the testing program has given rise, as an unexpected corollary, to the development of a strong network of mathematics teachers in New Mexico—a prerequisite for communication within a primarily rural state.

## Description of the New Mexico JUMP Testing Project

JUMP is an acronym for Junior Mathematics Prognosis, which is a testing program for high school juniors administered by the University of New Mexico, Department of Mathematics and Statistics. The New Mexico tests, as adapted from other similar programs, are designed to measure the students' readiness for college-level mathematics courses. The primary objective is to inform the students of any mathematical deficiencies, while they still have an opportunity to correct them, by taking the appropriate mathematics courses during their senior year of high school.

Two levels of tests have been developed by a committee composed of university and high school mathematics faculty. One level (Test A) is designed to predict college placement in remedial math, intermediate algebra, or college algebra. The scope of the other level (Test B) makes it possible to assess readiness skills for courses such as intermediate algebra, college algebra, trigonometry, and calculus. Within each of the tests, several content areas are evaluated (e.g., algebra, geometry, trigonometry, or precalculus).

In contrast to the Ohio Early College Mathematics Placement Testing Project, a JUMP student is given a separate subscore for each of the content areas. Students who take Test A receive three subscores: (a) arithmetic and introductory algebra, (b) geometry, and (c) intermediate algebra. Those students who take Test B receive two subscores: (a) algebra and precalculus and (b) geometry and trigonometry. The participating teachers agree that this manner of scoring provides a better diagnostic tool than a mere single composite score. A Sample Test B is an appendix to this article, as is a student response letter that was generated based on the test results.

The individual schools decide which juniors will take Test A and which juniors will take Test B. The testing takes place at the individual high schools during the months of February and March. No records are kept on individual student performance or overall school performance. Unlike other mathematics testing instruments, the JUMP test is not intended for the purpose of comparing scores or accumulating data on the students and schools; it does not measure general aptitudes and abilities in mathematics, and it lets the students know whether their preparation for postsecondary mathematics is adequate. Adequate preparation is based on an entrance examination and skills needed in the introductory collegiate mathematics courses of precalculus and calculus. This type of information can be a great advantage to the student, if it leads to corrective action. That is, the student still has an opportunity to seek the advice of teachers and counselors about enrolling in the appropriate mathematics class during the senior year of high school.

As the schools complete the actual testing, all answer sheets are returned to the UNM Testing Center where they are machine graded, and a personalized letter is generated for the individual student. This letter is used as a means of conveying the student's subscores in a personal, nonthreatening manner. It also provides an opportunity to list the college math courses that the student seems qualified to take. Specific UNM course numbers are not used. Instead, the suggested course titles are presented in general terms, which are consistent with those course titles found throughout New Mexico in postsecondary institutions.

Early indications are that the program has succeeded in giving students a realistic impression of the expectations that exist for college and university students in mathematics. This realization has resulted in increased enrollment in senior mathematics courses among some participating schools. Other schools do not report significant enrollment increases by seniors, but they do report seniors taking their mathematics course work more seriously after having taken the JUMP test. Considerable mathematics awareness also results from the test, as a brochure describing the test and the purpose is widely distributed and, in some cases, handed out at PTA meetings.

In 1989, a major news article went out on the New Mexico AP wire and was even picked up by the *Denver Post*, in addition to most New Mexico newspapers. This publicity helped to create school and community interest and led to many schools asking to be included. JUMP teachers have made presentations at statewide National Council of Teachers of Mathematics (NCTM) meetings, and mailings to all high schools have been used to help generate interest. Parent, student, and teacher response all seems to be positive. As a cost-effective project, it has far surpassed our expectations in beginning to change student and parent attitudes and enhance awareness of mathematics readiness.

## The Teacher Networking Corollary

New Mexico, a primarily rural state, is larger in area than the six New England states, New York, and New Jersey combined; yet it has a population of 1.5 million, which is less than that of metropolitan Seattle. Albuquerque is the largest city with over 500,000, while the rest of the population base is primarily stretched out along the Rio Grande River valley which runs down the middle of the state.

Because of the modest size of most of the secondary schools in New Mexico and the extreme distances between even medium-sized communities, many teachers work in isolated environments and find it difficult to maintain close contact with their colleagues and to share new ideas about teaching strategies, course content, and curriculum development. Moreover, students are also affected by this isolation and often lack a suitable context for academic comparison. Some JUMP coordinating teachers report that the nearest high school mathematics teacher is over sixty miles away, and they are fortunate to meet, except by chance, at school sporting events.

Realizing this situation, the JUMP program has made every effort to extend the benefits of the project beyond the knowledge gained by the high school juniors and to secure some advantages for the mathematics teachers and participating schools. With this objective in mind, we set out to foster linkages that would begin to develop a teacher network system—with high school mathematics teachers at the heart of that system. These linkages are briefly described in the sections that follow.

**1. High school math department.** Several teachers have reported that the process undergone during the testing project has prompted a more critical analysis of their own departments' mathematics curriculum. Given the small size of most of our rural high schools, teachers sometimes find it difficult to realistically evaluate "good" students, who are necessarily also a small population. Knowing something about what is expected of students in college math courses has helped the teachers to reflect back on the needs of students in high school math courses. This knowledge has enhanced the level of communication within the math departments, motivated primarily by an effort to adjust their expectations, and to more accurately gauge student accomplishment—especially when student populations are small.

**2. High school and middle school faculty.** In a few high schools, the mathematics teachers began to view themselves in a new light: as a bridge between middle school and college mathematics. Therefore, they began to develop linkages with their counterparts in the "feeder" middle schools. By opening up lines of communication, these high schools gained a perspective on where their mathematics students were coming from before high school and where they were going after high school. This knowledge has been instrumental in influencing changes in the mathematics programs in these particular high schools.

**3. High school math and counseling departments.** The JUMP process has nurtured a very strong internal linkage between each high school's mathematics department and counseling department. The counselors play a critical, complementary role in working with the mathematics teachers and using the test results to guide students into the most advantageous math courses for the senior year. In this way, the counselors have become more knowledgeable about the mathematics curriculum and more aware of the expectations at the high school and postsecondary levels.

**4. High school and university mathematics connection.** The creation and maintenance of a test-writing committee is an intentional first step in the development of constructive articulation between high school and college mathematics faculty. The test items selected are judged to be a representative sample of the concepts presently taught in New Mexico's high schools and thought necessary for success in courses offered in postsecondary institutions. By bringing together pieces of the curricular puzzle, university and high school faculty are able to assemble the interlocking parts and, consequently, provide a sense of continuity and relatedness across mathematics courses.

Direct contact with the participating high schools is a second step in maintaining constructive articulation between high school and college mathematics faculty. In the New Mexico JUMP model, this is accomplished through two general meetings which are held at the UNM campus. The first meeting—Orientation Meeting—takes place during the month of January, and the second meeting—Evaluation Meeting—takes place during the latter part of April. Each participating school makes a commitment to send two representatives to the general meetings: one representative is required to be a high school math teacher, and the other must be a high school counselor. In fact, at the two annual JUMP coordinator meetings, we usually have several principals or superintendents show up with their mathematics teacher to discuss the program and interact with the other meeting participants.

These meetings have made it possible to share information about the teaching and learning of mathematics and about the structure of the mathematics curriculum. The information acquired as a result of the verbal group discussions and written evaluation forms is carefully analyzed and provides a basis for further modification of the JUMP program. In this way, the schools play a genuine and vital role in the entire testing process. We feel that this unique aspect of the New Mexico model is the key to the teacher networking advantage and to the enthusiastic response of teachers and schools.

## Concluding Remarks

The New Mexico model shows evidence that the benefits of prognostic testing can go well beyond the message delivered to the junior high school students. In a primarily rural state such as New Mexico, JUMP has provided an excellent vehicle for the growth of a teacher network. Such a network is a critical factor in the development of a sense of professionalism in mathematics teachers—especially those who are working in remote, isolated conditions.

Because of a lack of opportunities to meet with colleagues, the teachers have enthusiastically supported JUMP as a program which has succeeded in providing an outlet whereby they are able to share ideas and experiences about the teaching and learning of mathematics. According to one teacher, the JUMP project is "already a success in that it has brought math teachers from all over the state together, allowed them to get acquainted, and encouraged them to discuss the problems of math education."

## Appendix

# NEW MEXICO JUMP PROJECT
The University of New Mexico
Department of Mathematics & Statistics

Dear John Doe:

This is a report on how you did on the JUMP Test you took a few weeks ago. The JUMP Test is designed to help you evaluate your progress in preparing for college level courses in mathematics. This test is similar to college placement tests that you will take when you enroll at a college or university.

Your results on the JUMP Test Form B are as follows:

*Algebra & Precalculus*      11 out of 20
*Geometry & Trigonometry*    05 out of 10

Consider the chart on the back of this letter. It will help to interpret your score and shows your suggested placement if you were to apply now to a university for admission.

The scores that you received on this test and their evaluations are <u>indications</u> of your progress and are not absolute. If you missed everything in a particular content area, then you have a problem. If you received a perfect score, then you probably have a solid foundation in that specific content area. On the other hand, if you are weak in an area where you feel you know the material, or you are satisfactory in an area in which you have doubts, or your scores are ambiguous, then you should discuss the results with your teacher, your counselor, and your parents.

In all cases, if you plan to go to college, you should take a mathematics course in your senior year. The loss of skills incurred by not taking math in your senior year of high school may put you at a serious disadvantage when you enter college or any other postsecondary educational institution.

Most fields of specialization in college require a number of college level math courses. If you are interested in learning more about math requirements for various subject majors, see your high school math teacher or counselor. Also, you should ask them about which senior level math course would be most beneficial to you. Good luck in your future educational plans!

Sincerely,

Nancy Gonzales, Director
New Mexico JUMP Project

# WHAT DO THE SCORES MEAN?

The sequence of courses that are referred to below are more or less standard at most colleges:

1. *Arithmetic and introductory algebra* (remedial and often not offered at colleges).
2. *Intermediate algebra*, which is roughly equivalent to high school Algebra I & II (remedial and if it is offered at college you might not receive credit for it).
3. *College algebra*, which is the course that leads to calculus.

Knowledge of geometry is usually assumed at the college level and there are no courses that cover this material.

Consider the chart below which will help to interpret your score and which shows your suggested placement if you were to apply now to a university or college for admission.

| CONTENT AREA | SCORE | COLLEGE PLACEMENT/EVALUATION |
|---|---|---|
| Algebra & Precalculus -Problems exemplifying algebra skills essential for calculus that should be known in the junior year. Further essential skills will be acquired in your senior year. | 0 – 6 | Intermediate algebra. One course below college level (or lower). Often not offered for college credit. A senior year mathematics course is strongly recommended. |
|  | 7 – 14 | College algebra (if you continue your progress in a senior year mathematics course). At this point, you seem to be making substantial progress! |
|  | 15 – 20 | Very good progress! Continuing in your senior year, your skills should be sufficient for calculus. Keep up the good work. |
| Geometry & Trigonometry -This section deals with the trigonometry for calculus and the geometric knowledge which is assumed in calculus. | 0 – 3 | Weak. Geometry skills need to be addressed before taking trigonometry. |
|  | 4 – 6 | Trigonometry (or precalculus if offered). |
|  | 7 – 10 | Great progress! If your algebra and precalculus skills are also strong, with continued progress as a senior you should not have to take trigonometry before (or with) calculus. |

# FORM B

### JUMP

### NEW MEXICO JUNIOR MATHEMATICS PROGNOSIS TEST

### UNIVERSITY OF NEW MEXICO
### DEPARTMENT OF MATHEMATICS AND STATISTICS

Spring 1991
Form B

INSTRUCTIONS AND INFORMATION

1. DO NOT OPEN THIS BOOKLET UNTIL TOLD TO DO SO BY YOUR INSTRUCTOR.

2. This test contains thirty multiple choice questions. Each question is followed by five possible answers labeled (a), (b), (c), (d) and (e). Only one answer is correct.

3. For each question, indicate your answer by marking the appropriate space on the answer sheet provided by your instructor.

4. Your score on this test is the number of correct answers. However, we ask that you avoid random guesses because this will hinder our ability to provide a correct diagnosis of your mathematical skills.

5. Use a #2 pencil to mark your answers on the answer sheet. Scratch paper and erasers are permitted. Calculators are not permitted.

6. The figures provided in the problems are not necessarily drawn to scale.

7. Before beginning the test, print and code in the following information on the answer sheet: name and social security number , and answer the information questions.

8. Answer as many questions as you can in the time allotted, but don't spend too much time on any one question. You can expect to see some problems which may not be familiar to you. Don't become discouraged; just do the best you can without random guessing.

9. PLEASE DO NOT WRITE ON THIS TEST BOOKLET.

1. By rationalizing the denominator, the expression $\dfrac{\sqrt{5}}{3-\sqrt{5}}$ simplifies to:

    (a) $\dfrac{1}{2}$    (b) $\dfrac{3\sqrt{5}-5}{11}$    (c) $\dfrac{3\sqrt{5}+5}{4}$    (d) $\dfrac{3\sqrt{5}+5}{14}$    (e) $\dfrac{3\sqrt{5}-5}{4}$

2. If $V = P(1+rt)$, then $t$ is equal to:

    (a) $\dfrac{V}{rP} - 1$    (b) $\dfrac{V-P}{r}$    (c) $\dfrac{V-1}{rP}$    (d) $\dfrac{V-P}{rP}$    (e) $\dfrac{V+P}{rP}$

3. The $x$-coordinate of the solution of the system of equations $\begin{cases} 2x + y = 3 \\ x - 2y = 4 \end{cases}$ is:

    (a) $-1$    (b) $\dfrac{2}{3}$    (c) $2$    (d) $\dfrac{7}{3}$    (e) $4$

4. $\sqrt[3]{\dfrac{x^{3/2}y}{y^{-2}}} =$

    (a) $x^{9/2}y^9$    (b) $x^{1/2}y$    (c) $x^{9/2}y^3$    (d) $x^{1/2}x^{1/3}$    (e) $x^{1/2}y^3$

5. The volume, $V$, of a cylindrical can whose radius is $r$ and whose height is $h$ is given by $V = \pi r^2 h$. The surface area, $S$, of the can is given by $S = 2\pi rh + 2\pi r^2$. If the volume of the can is $V = 24$ cubic inches and the radius of the can is $r = 2$ in, what is the surface area (in square inches)?

    (a) $16\pi$    (b) $8 + 24\pi$    (c) $32\pi$    (d) $24 + 8\pi$    (e) $32$

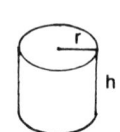

6. The graph of $7x + y + 21 = 0$ crosses the $x$-axis at $x =$
    (a) 21    (b) 3    (c) 7    (d) $-3$    (e) $-7$

7. The equation of a line containing the points $(2, -1)$ and $(-3, 2)$ is:

    (a) $5x - 3y = 11$    (b) $3x - 5y = 1$    (c) $3x + 5y = 1$    (d) $5x + 3y = 1$    (e) $3x - 5y = 11$

8. $\dfrac{2x}{x^2-1} - \dfrac{1}{x-1} =$

   (a) $\dfrac{1}{x-1}$ (b) $\dfrac{1}{x+1}$ (c) $\dfrac{2x-1}{x^2-1}$ (d) $x-1$ (e) $x^2-2x+1$.

9. $\dfrac{3a}{3+6a^2} =$

   (a) $\dfrac{1}{3a}$ (b) $\dfrac{1}{2a}$ (c) $\dfrac{1}{1+6a^2}$ (d) $\dfrac{1}{1+2a}$ (e) $\dfrac{a}{1+2a^2}$

10. The solution set of $3x^2 + x = 4$ is

    (a) $\{1\}$ (b) $\left\{-\dfrac{4}{3}\right\}$ (c) $\left\{0, -\dfrac{1}{3}\right\}$ (d) $\left\{1, -\dfrac{1}{3}\right\}$ (e) $\left\{1, -\dfrac{4}{3}\right\}$

11. The graph representing $|x-4| \geq 2$ is:

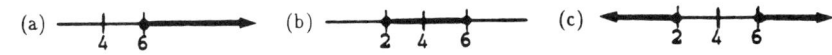

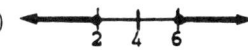

12. If $f(x) = 2x^2 - 3x$, then $f(1-h)$ is equal to:

    (a) $2h^2 - 7h - 1$ (b) $2h^2 - h - 1$ (c) $-2h^2 + 3h - 1$ (d) $2h^2 - h + 5$ (e) $2h^2 - 7h + 5$

13. The graph of a function $y = f(x)$ is shown. Exactly one of the following statements is false. It is:

(a) $f(1) = 0$  (b) $f(-1) > f(0)$  (c) $f(-2) > 0$  (d) $f(2) < 0$  (e) $f(-3) < f(3)$

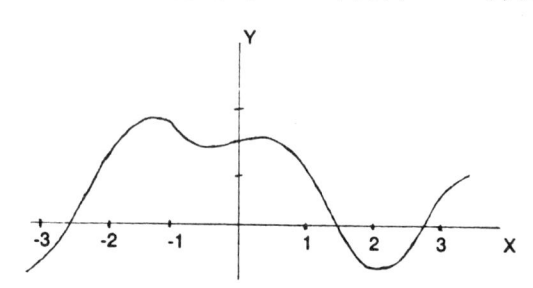

---

14. If $f(x) = x^2 - x$ and $g(x) = 1 - x$, then $f[g(x)]$ equals:

(a) $x^2 - 3x$  (b) $-x + 1$  (c) $x^2 - x$  (d) $-x^2 - x$  (e) $-x^2 + x + 1$

---

15. If $25^{3x} = 5$, then $x$ is equal to:

(a) $\frac{1}{3}$  (b) 0  (c) $-\frac{1}{2}$  (d) $\frac{1}{6}$  (e) No solution

---

16. The points of intersection of the line $x + y = 6$ and the parabola $y = x^2 + 4x$ are:

(a) $(-6, 12)$ and $(1, 5)$  (b) $(-3, 9)$ and $(-2, 8)$  (c) $(-1, 7)$ and $(6, 0)$  (d) $(2, 4)$ and $(3, 3)$

(e) They do not intersect.

---

17. If $x^3 + 2x^2 + 1$ is divided by $x^2 + 3$, then the remainder is:
(a) 2  (b) $-3x - 5$  (c) $-x + 1$  (d) $3x - 5$  (e) $-3x + 5$

---

18. Given that $x = 1$ is a root of the equation $x^3 - 2x^2 + 3x - 2$, the other roots are:

(a) $-1$ and 2  (b) $\frac{1 \pm i\sqrt{7}}{2}$  (c) 1 and $-2$  (d) 0 and 3  (e) $\frac{3 \pm \sqrt{17}}{2}$

19. The graph of $y^2 = 1 - x^2$ is:

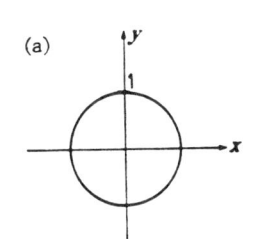

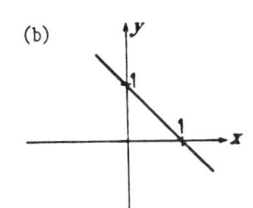

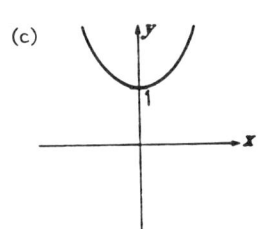

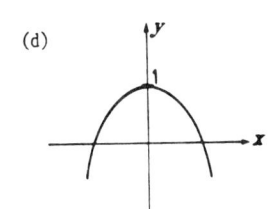

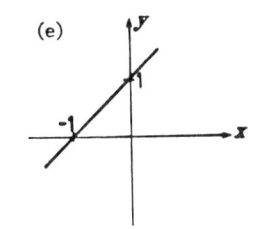

(a)  (b)  (c)

(d)  (e)

20. If $x > 0$, then $\log\left[\dfrac{x+1}{x}\right]$ equals:

   (a) $\log \dfrac{1}{x}$   (b) $\log(x+1) + \log x$   (c) $\log(x+1) - \log x$   (d) $\dfrac{\log(x+1)}{\log x}$   (e) $\log 1$

21. The figure below is a square in which $\overline{AB}$ has length 1 and $\overline{AC}$ has length 2. The area of the square is:

   (a) $\sqrt{3}$   (b) $\sqrt{5}$   (c) 3   (d) 4   (e) 5

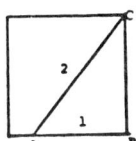

22. In the figure, if $BE = BD = DC$, then the degree measure of the angle $x$ is:

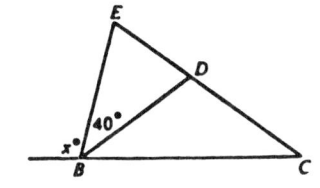

   (a) 35   (b) 75   (c) 105   (d) 115   (e) 120

23. In the figure at the right, $\overline{RS}$ and $\overline{QT}$ are altitudes and $\overline{PS} \cong \overline{PT}$.
Which postulate or theorem should be used to prove $\triangle RSP \cong \triangle QTP$?
(a) SSS   (b) ASA   (c) SAS   (d) AAA   (e) Hypotenuse - Leg

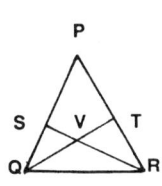

24. A playing field is to be built in the shape of a square of side $x$ plus a semicircular area at each end. What is the perimeter of the field in terms of $x$?
(a) $4x$   (b) $2x + \pi x$   (c) $2x\pi$   (d) $x + 2\pi x$   (e) $\pi x$

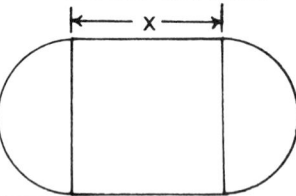

25. The length of the circular arc AB centered at the origin in the drawing is:
(a) $\pi$   (b) $2\pi$   (c) $\dfrac{\pi}{2}$   (d) $\dfrac{\pi}{3}$   (e) $\dfrac{\pi}{4}$

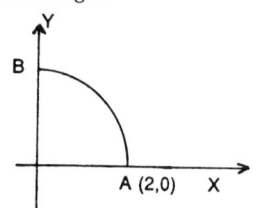

26. In the triangle at the right, the value of $\cos A$ is:
(a) $\dfrac{4}{5}$   (b) $\dfrac{5}{3}$   (c) $\dfrac{5}{4}$   (d) $\dfrac{3}{4}$   (e) $\dfrac{3}{5}$

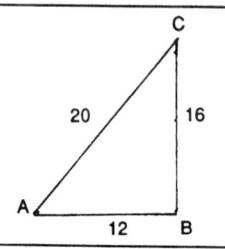

27. The radian measure of the angle $\theta$ shown in the drawing is:
(a) $\pi$   (b) $\dfrac{5\pi}{2}$   (c) $3\pi$   (d) $6\pi$   (e) $\dfrac{3\pi}{2}$

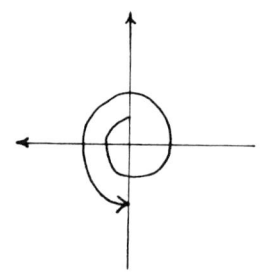

28. If $\theta$ is an angle in the 4th quadrant and $\cos\theta = \frac{4}{5}$, then $\tan\theta =$

(a) $\frac{5}{3}$   (b) $\frac{4}{5}$   (c) $-\frac{3}{4}$   (d) $-\frac{4}{3}$   (e) $-\frac{4}{5}$

---

29. What is the largest angle $x$, where $0 \leq x \leq \pi$, such that $\frac{1}{2} - \sin 2x = 0$?

(a) $\frac{\pi}{6}$   (b) $\frac{\pi}{3}$   (c) $\frac{\pi}{12}$   (d) $\frac{5\pi}{12}$   (e) $\frac{5\pi}{6}$

---

30. One cycle of the graph of $y = \cos\left(\frac{x}{2}\right)$ is:

(a)

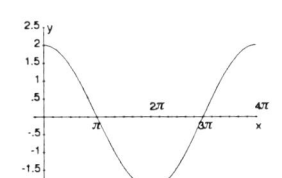

(b)

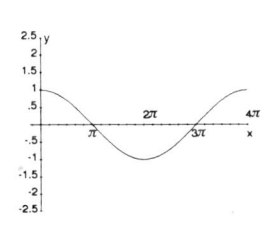

(c)

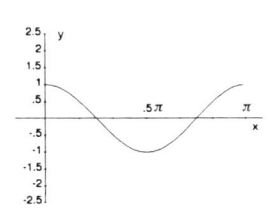

(d)

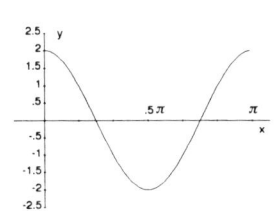

(e)

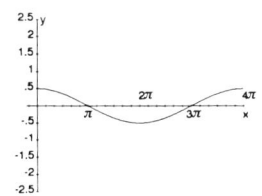

## References

1. F. Gilfeather, N. Gonzales, and D. Miller, *Establishing pilot junior prognostic tests*, Proceedings of a National Conference on Prognostic and Diagnostic Testing in Mathematics, Washington, DC, November 1988; also MAA Monograph (to appear).

DEPARTMENT OF MATHEMATICS AND STATISTICS, UNIVERSITY OF NEW MEXICO, ALBUQUERQUE, NEW MEXICO 87131

# Using Technology for Teaching Mathematics

CARL SWENSON

## Introduction

Although I have been a computer user for twenty-five years, it is only recently that I have started to use the computer as a tool for teaching mathematics. In my case, having overhead projection capabilities, computer labs, and adequate hardware/software enables me to improve my teaching so that the results do justify the efforts to use the technology.

My experiences with technology in the classroom have taught me some important practical lessons which I would like to share with other mathematicians, especially those who are just beginning to use computers and other technology in their teaching. My focus is on well-established examples that can be comfortably tackled by a single teacher or by a small group within a department. These suggestions begin with some straightforward recommendations and move to questions one might consider in using a sophisticated program such as *Mathematica*. I hope they will help other mathematicians select the technology and software that suit their needs and also enable them to avoid some potential pitfalls of using technology in teaching.

## Using Technology in Teaching

There are three main categories of technology used in teaching mathematics. First, you can use a personal computer for the preparation of course materials and administrative tasks; second, you can do projected screen presentations in the classroom; and third, you can teach in a computer lab. I suggest that you proceed in that order and first invest the time to learn a word processor and/or learn a spreadsheet if you do budgets or keep numeric data. Choose a word processor that handles mathematical notation. Good general word-processing choices are WordPerfect 5.1 for the IBM, which has an equation editor, and Word 4 for the Macintosh, which has the ability to write most common symbols and formats. Several specialized mathematical editors are also available; these are especially attractive if your work involves only mathematical writing. As you become more advanced, a TeX product will give you complete notational and format control for journal articles and publications.

My reluctance to use a personal computer in the classroom was overcome when I had control of the screen projection. There were too many disasters when I had to specially schedule a presentation in a room with a stationary projection system. Now screen projection panels sit on an overhead projector and the presentation can be from your own computer. It still happens that doing something for the first time by computer takes longer than the previous, by hand, methods, but efforts in selecting software and gaining proficiency in using the software will pay you back in the long run. The benefits are discussed in a later section.

Computer labs are not simple. Networks are often used and they require a high level of expertise to set up and maintain. Software must be protected and default settings maintained from one student use to the next. Before your students begin work in an existing lab, use it yourself several times and observe how it functions. If the equipment is not identical, test each model to insure software compatibility. Before creating a new lab, be sure to visit and talk to directors of other labs that are functioning.

## GETTING STARTED

**1. Start with simple tasks.** Start with simple software that allows you to learn something about your computer while you master the software. For example, packages like MasterGrapher, or True Basic's Math Series (Calculus, etc.) are worthwhile and yet relatively easy to use. A graphing calculator, such as the TI-81 or the HP48sx is an appropriate choice of technology for most undergraduate courses. (See an excellent review of graphing calculators by Demana, et al., in *The Computing Teacher*, April 1990.) Because of the low cost and convenience of graphing calculators they can be realistically required and used in class activities and exams. These powerful calculators have projected screens available and are a safe way to begin using projected technology.

**2. Find support around you.** Unless you are a brave-hearted Lone Ranger with very special needs, the other computer users around you are one of the most important, but often least considered, assets in helping you make hardware/software choices. Contacts can be made through your school computer support services, or a local users' group, or a knowledgeable colleague. Magazine recommendations can be misleading and toll-free product help can keep you on hold and be unreliable. Begin in your local mainstream. National support is available through the AMS/MAA meetings where minisessions are predominantly computer related. Ohio State University sponsors the annual International Conference on Technology in Collegiate Mathematics.

The MAA has published *Computers and Mathematics: The Use of Computers in Undergraduate Instruction* and the *Notices of the American Mathematical Society* has a monthly column entitled "Computers and Mathematics."

**3. Ask for support.** Technology is a hot item; administrators do not want their faculty perceived as still being in the middle ages. Administrators can and should be easily convinced that the use of technology in education is time-consuming and requires faculty development support, such as release time, sabbaticals, and support for attending conferences and minisessions.

**4. Guard your time.** You are wasting time if you try out software—especially free offers—unless it meets your own specific and pressing needs. Why spend (lose) two days playing with a program, even a free one, that puts passwords on your files when you are the only user of your computer and your office door has a lock? Whenever possible avoid getting involved in setting up computer connections or initializations; use technical support.

**5. Build on success.** Start with relatively fail-safe presentations in the classroom. Practice in advance using an overhead projected monitor or whatever equipment you will actually use in the classroom. It is very frustrating to waste class time trying to make equipment work correctly. Do not make assumptions about projection device compatibility. While it is obvious that an IBM projection device may not accommodate a Macintosh, even the Macintosh SE and Macintosh II series can differ in connections. The safest arrangement is to have a dedicated machine equipped with projection capability. Check the projection settings and have the computer running before class or else set it up in class, during a mini-quiz or group work time.

If you get off the path and obtain unexplainable results during a presentation, stop and promise to clear up the confusion at the next session. Be careful about going too fast—this is the lesson of all first-time camcorder video makers who whirl around to show the landscape in one blurring sweep. Learn in your home or office; make classroom presentations before embarking on any full-scale projects that involve a computer lab for a whole class.

My recommendation is to keep in the mainstream. This is a warning: pioneering is expensive, time-consuming, and dangerous.

## THE NEXT STEP

The next step, given that you have a good purpose and some experience, is to move to a higher level of sophistication. You may have to branch out to new hardware or software. To efficiently incorporate more sophisticated technology in your teaching, I suggest that you look at the following categories: power and ease-of-use, mathematical notation and presentation, portability, support, and cost. Below, I use *Mathematica*, a commercial program advertised as a system for doing mathematics, to illustrate how you might evaluate a particular software package according to these categories. Scientists and engineers are beginning to use *Mathematica*; for this reason alone I feel that the mathematics community must keep an eye on its development. I recently overheard a discussion among engineers in which it was hoped that *Mathematica* would allow students to take fewer mathematics courses, as students could be taught to use *Mathematica* for the mathematics needed in engineering courses.

## Overview of Mathematica

Some background information about *Mathematica* is useful for following the illustration. *Mathematica* incorporates numeric computation, symbolic manipulation (algebra and calculus), two- and three-dimensional graphing, list processing, and a powerful programming language. If you do not know or do not like programming, you may not want to use *Mathematica* in its present form.

Notebooks in *Mathematica* are similar to documents or files for a word-processor; they are the workspace. In a notebook, the user types and edits commands, and/or text, which can be sent to a "kernel" for evaluation. The result of the evaluation, either text or graphics, is output on the next lines of the notebook. The kernel stores all of its calculations during any one session so they can be reused. Example 1 shows a notebook with three examples, each containing a heading text comment, a command in the *Mathematica* language, and the kernel's evaluation output. Usually, many examples having a theme or learning sequence are combined to create a single notebook.

All work, whether text or graphics, is put into a cell and has square demarcation brackets on the right-hand side of the notebook. A notebook is more than a simple file because the bracketing is hierarchical which, like an outline, can be collapsed or expanded. Examples 1 and 2 show the brackets as they appear on the screen. Example 2 is a partially collapsed version of example 1; the first section is the same in both, the second section has the output cell closed (hidden), and the third section has both the input and the output cell closed. This and other notebook editing features make *Mathematica* convenient for presentations or for writing notebooks that allow students to be guided by prewritten commands while still having the ability to try out their own examples.

## Software Evaluation

While your mathematical purpose should drive your software selection, the following considerations can also be helpful in evaluation.

**Power versus ease of use**. The most difficult balance to achieve in software is power versus ease of use. A dilemma of software design is that if it is powerful and all-encompassing then it is also more difficult to use and the potential for error much greater. *Mathematica* requires a substantial commitment of

■ **Numeric precision**

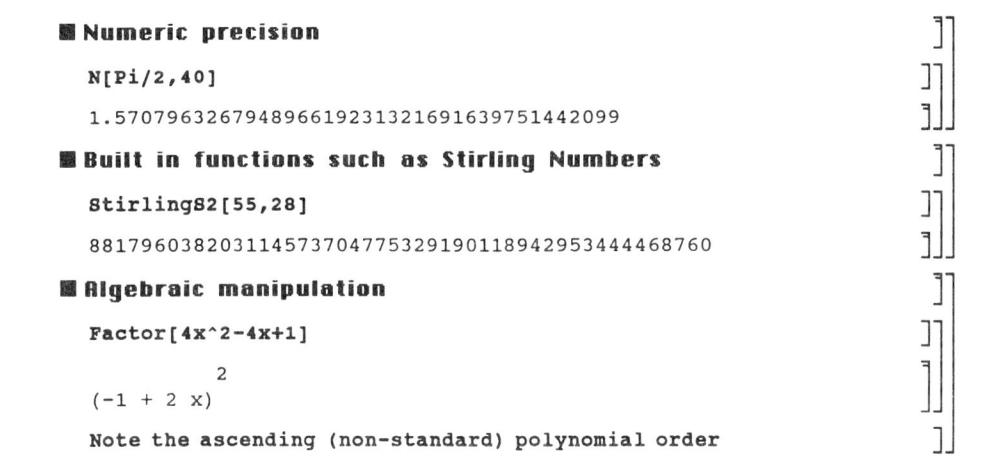

    N[Pi/2,40]

    1.5707963267948966192313216916397514420
99

■ **Built in functions such as Stirling Numbers**

    StirlingS2[55,28]

    881796038203114573704775329190118942953444468760

■ **Algebraic manipulation**

    Factor[4x^2-4x+1]

                 2
    (-1 + 2 x)

    Note the ascending (non-standard) polynomial order

EXAMPLE 1

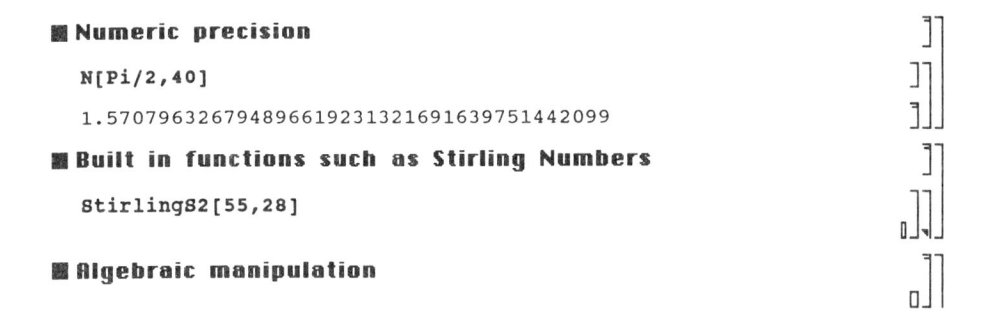

■ **Numeric precision**

    N[Pi/2,40]

    1.5707963267948966192313216916397514420
99

■ **Built in functions such as Stirling Numbers**

    StirlingS2[55,28]

■ **Algebraic manipulation**

EXAMPLE 2

effort and attention—often subtle differences such as capitalization, spaces, or representation (needing to use 1/2 instead of 0.5, for example) can frustrate the user. See Example 3 (see p. 166).

At this time, *Mathematica* 1.2 is not as attractive to teachers as specialized programs which offer ease of use in elementary situations. At every presentation I have given on *Mathematica*, some one has come up to me and said, "I use (Derive, Theorist, MasterGrapher, ...) and it's much easier." The developers of *Mathematica* are betting that in the end, users will opt for a powerful, comprehensive, flexible, programmable, single environment.

## ∎ Difference in powers .5 versus 1/2

```
Integrate[ (x^(1/2) - 1)/(x - 2), x]        (* This is great!! *)
                    2 Sqrt[2] - 2 Sqrt[x]
        2 Log[---------------------]
                    2 Sqrt[2] + 2 Sqrt[x]
2 Sqrt[x] + ------------------------------ - Log[-2 + x]
                        Sqrt[2]
Integrate[ (x^(.5) - 1)/(x - 2), x]         (* Can't do this one!! *)
               0.5
              -1 + x
Integrate[---------, x]
              -2 + x
```

EXAMPLE 3

**Mathematical notation and presentation.** The next issue, and one that is somewhat unique to mathematics, is how *Mathematica* handles mathematical notation and formats. Since notation within *Mathematica* cannot be ambiguous as it is in mathematics, some deviation from standard mathematical notation is required. For example, parenthesis pairs must have a single meaning—not the six to ten different uses in mathematics! Thus, *Mathematica* interprets $3(x + 2) = 3x + 6$ but avoids ambiguity with functional notation by using $f[\ ]$. Different meanings of "equals" are distinguished as: assignment ($x = 3$, $x$ is immediately assigned the value 3); set delayed ($f[x\_] := 3x$, later $f[5]$ will be evaluated as 15); and logical testing ($x == 3$, has value true or false.) As an example of the power of the notation, Example 4 shows how you can create your own recursively defined function. Having to use nonstandard notation of *Mathematica* has made me more empathetic with mathematics students as they struggle with ambiguous, contextually-sensitive notation.

Sooner or later you will need to confront the problem—or joy—of fonts. A font is a set of defined graphic representations for the standard characters that appear on the screen and printer. In Example 1, the section titles use Chicago font, while the input and output cells use Courier font. In addition, each font has styles (**bold**, underline, *italics*, ...); in Example 1, the input cells are bold and output cells are plain. Different font sizes can be used; for example, material created for a small classroom, but later reused in a large lecture hall, might require that you make the size of the font larger. Remember meetings with typed overheads that could not be read in the second row? In *Mathematica* it is possible to globally or locally change the font, size, color, or style of text at any time. Using more than one font is an effective way to clarify different aspects of the material.

Notebooks are well suited for authoring student lessons; and while they are available on the Macintosh, Next, and IBM 386 (through Windows), they

■ **Defining the Fibinacci function recursively:**

```
fib[1] = 1; fib[2] = 1; fib[n_] := fib[n-1] + fib[n
fib[8]
21
```

Note: ":=" is delayed set equal and "n_" is dummy variable.

EXAMPLE 4

are not implemented on workstations. A serious impediment of the present version is the inability to write mathematics in its traditional notation. When notebooks have been developed in mathematics, specialty fonts have been used to accommodate the special symbols. Recently I evaluated notebooks from three university projects, and I needed four different fonts to look at the work. While nonstandard fonts can be included on a disk and installed, in general they hinder portability, the next consideration.

**Portability.** Software available on only one machine can be a severe limitation in the long run as you start to share and even encourage diverse hardware in your department. There are different levels of portability. *Mathematica* has a common kernel that can be installed on the Macintosh, IBM 386, Next, Sun, DEC, and many workstations and minicomputers. This portability is excellent, but keep in mind the problem addressed earlier—"front end" and notebook capability differs from machine to machine. Another example is WordPerfect which is available on many different machines but only has an equation editor on the DOS machines. Another aspect of portability is the ability to share files among different applications; good software includes the capability to import and export files with many varied formats.

True Basic shows good portability between the Macintosh and IBM. This is the second generation of the famous procedural programming language BASIC, developed at Dartmouth College, with the intent of giving students an efficient tool to do powerful computer applications. True Basic has been developed with mathematics education in mind, and it has been used to write several easy-to-use stand-alone packages for undergraduate mathematics.

**Support.** As previously mentioned, you should choose software and hardware for which there is local help. However, as you become a more sophisticated user, you may need to look to the national level for conferences, publications, and user groups.

*Mathematica*, for example, has an annual international conference attended by people from a broad range of disciplines. Attendance by discipline at the 1991 meeting was reported as 28% Engineering, 21% Computer Science, 20% Physical Sciences, 12% Mathematical Sciences, 6% Life Science, and 6% Business. Some of the presentations were related to undergraduate mathematics education. A presentation at the high school level was made by

Sandra Dawson, the editor of the *High School Mathematica Network Newsletter*, from Glenbrook South High School, Glenview, Illinois.

Finding reference books and journals becomes more difficult as you begin using more specialized software. Good software spawns publications. There are a growing number of books available on *Mathematica* usage; all of the following references happen to be published by Addison-Wesley. The bible of *Mathematica* users is Wolfram's book, *Mathematica, A System for Doing Mathematics by Computer, 2nd ed*. Jerry Uhl and Horacio Porta from the University of Illinois at Urbana-Champaign produced *Calculus & Mathematica*, notebooks and printed material that completely replace the lecture-style course with work in *Mathematica* labs. An interesting recreational book, *Exploring Mathematics with Mathematica*, by Jerry Glynn and Theo Gray, includes a CD-ROM version of the book. (The arrival of CD-ROM and videodisc is one reason why I have used the word technology rather than computer.) If you become a regular *Mathematica* user and want to know what others are doing with *Mathematica*, *The Mathematica Journal* is a special-interest publication that includes the broadest range of applications and developments.

**Cost.** In considering your cost as a single user think about what your time is worth. In the long run, poor software will cost you a fortune! Thus my personal bias is to disregard cost for personal items; however, the cost for a computer lab can usually not be disregarded.

When planning to buy hardware and software, consider them together as a system. Buying only the computer first may mean later costs for more memory or a different graphics display that is demanded by your software. Printers and screen projection can often be shared but should be part of your plan. Take advantage of educational pricing and apply for grants. Lab equipment is becoming a standard part of mathematics instruction; thus grants, such as the National Science Foundation's Instrumentation and Laboratory Improvement (ILI) are available.

Using *Mathematica* as an example, the combined cost of hardware and software to run *Mathematica* is anywhere from $4,000 to $8,000. But if suitable computer labs already exist, student versions of *Mathematica* are only $139. An affordable option is the graphing calculator which now sells for between $80 and $300.

## QUESTIONS TO ASK YOURSELF

Here are some questions to keep in mind as you think about using technology in teaching mathematics.

**Is there a mathematical purpose driving the use of technology?** It is easy to get caught up in the razzle-dazzle of technology and let the use of technology become an end in itself. It is fun to create colorful, animated presentations, but they can distract from a concerted effort to improve your mathematics instruction. For example, the Sine curve can be animated to oscillate on a color screen, but does that have an impact on how students think about the Sine function? It is important to be honest with yourself: Are you introducing technology for its own sake or for the appearance of being modern, or is it to enhance your teaching of mathematics and open up new possibilities for your students?

**How will the use of technology help the learning of mathematics?** Answering this question in detail may save you an enormous amount of time and energy. Some things to consider are topics that can be enhanced through quality graphics presentations. For example, Figures 1 and 2 (see pp. 170 and 172) show the frames of an animation that I developed for a Calculus class. The graphics can be animated or shown as a slide show, and instant replay is available. In addition, the frames can be printed as shown in a single worksheet, then cut and stapled to form a flipbook. This seems to me a justifiable purpose because it is impossible to animate a chalkboard.

**Are there new ways to teach using the technology?** Simply translating the current way of teaching to a technological version will probably not produce much gain; for example, early television used a fixed camera to make educational films of famous mathematicians lecturing at the blackboard. The evolutionary aspect of technology has a feedback loop: when we change the tools available, we need to change the type of questions we ask and the kind of answers we expect. Some students believe that systems of equations always have integer solutions—I refer to these in class as "Hollywood answers." With calculators and computers available we can introduce real applications because the systems of equations that students can solve have been broadened.

Real breakthroughs could occur by using situations that require the students to conjecture, test, and write about mathematics. Just as science and engineering students go into labs and construct experiments or create models, it is possible to quickly set up a computer lesson in which students look at curve fitting by polynomials of successively higher orders. The polynomial coefficients are calculated and the resulting polynomial equations are easily plotted. Building, testing, tinkering, guessing, failing, and succeeding are all experiences that students can have with computer learning.

**What should we be teaching?** Given that *Mathematica*, or any other Computer Algebra System (CAS), can differentiate, factor polynomials, find roots, etc., what does this say about what we should be teaching? An initial reaction might be to forget computation and stick to the conceptual aspects of mathematics. However, even with full CAS support, students still need to learn to do simple examples by hand, which allows them a way to test the computer results before feeding it a complex example. I have observed—without

# The Fundamental Theorem Of Calculus

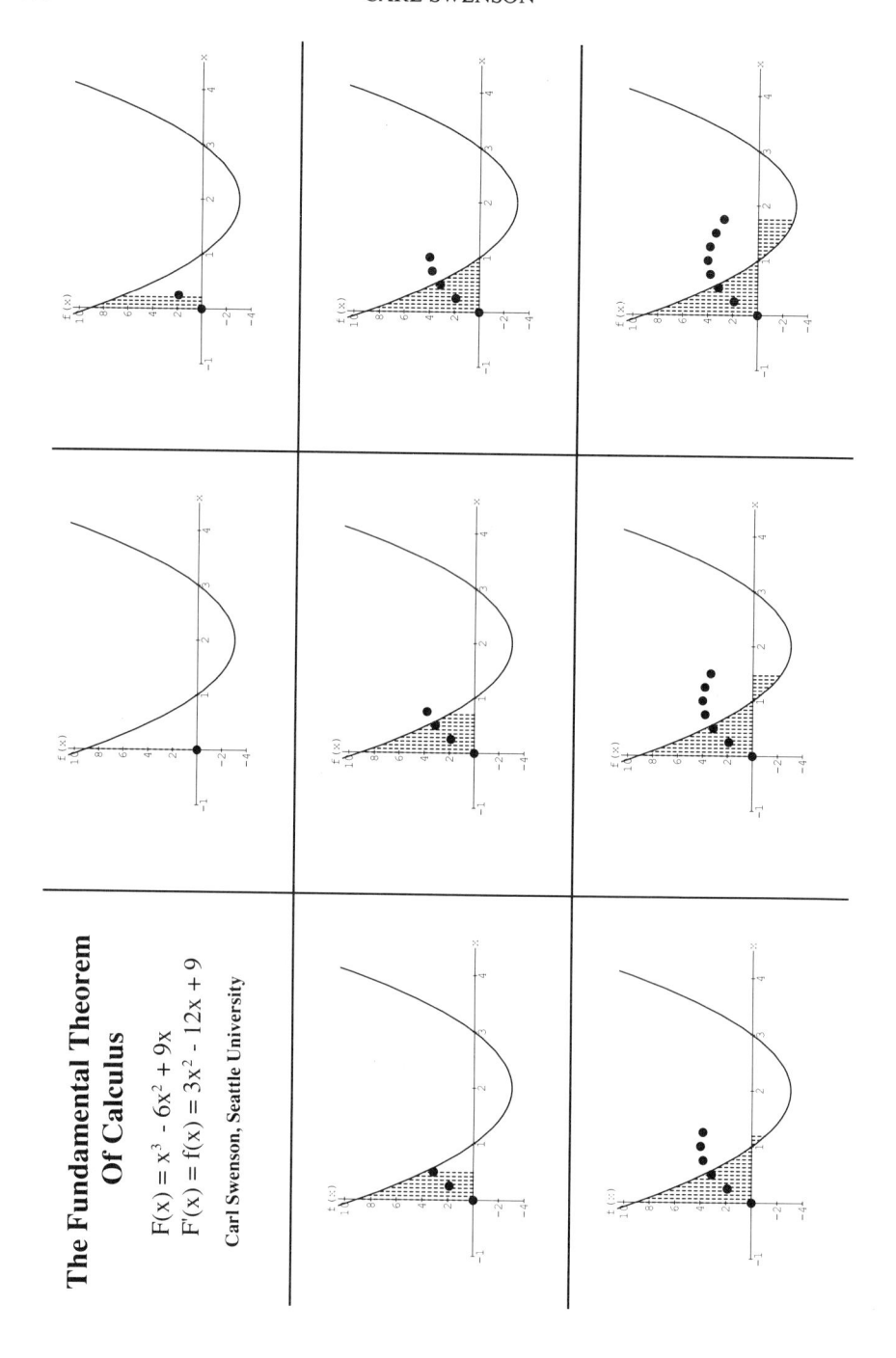

$F(x) = x^3 - 6x^2 + 9x$
$F'(x) = f(x) = 3x^2 - 12x + 9$

Carl Swenson, Seattle University

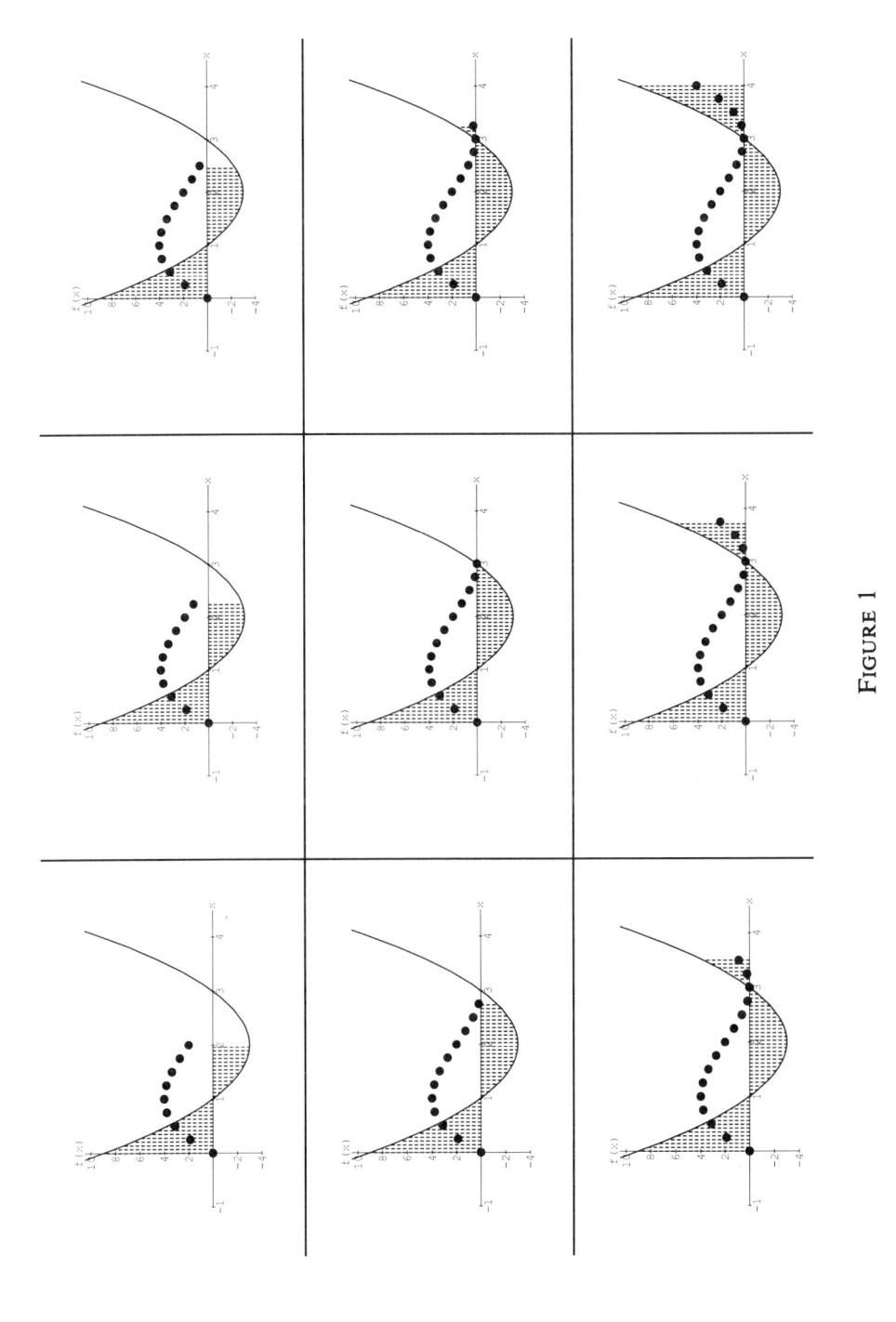

FIGURE 1

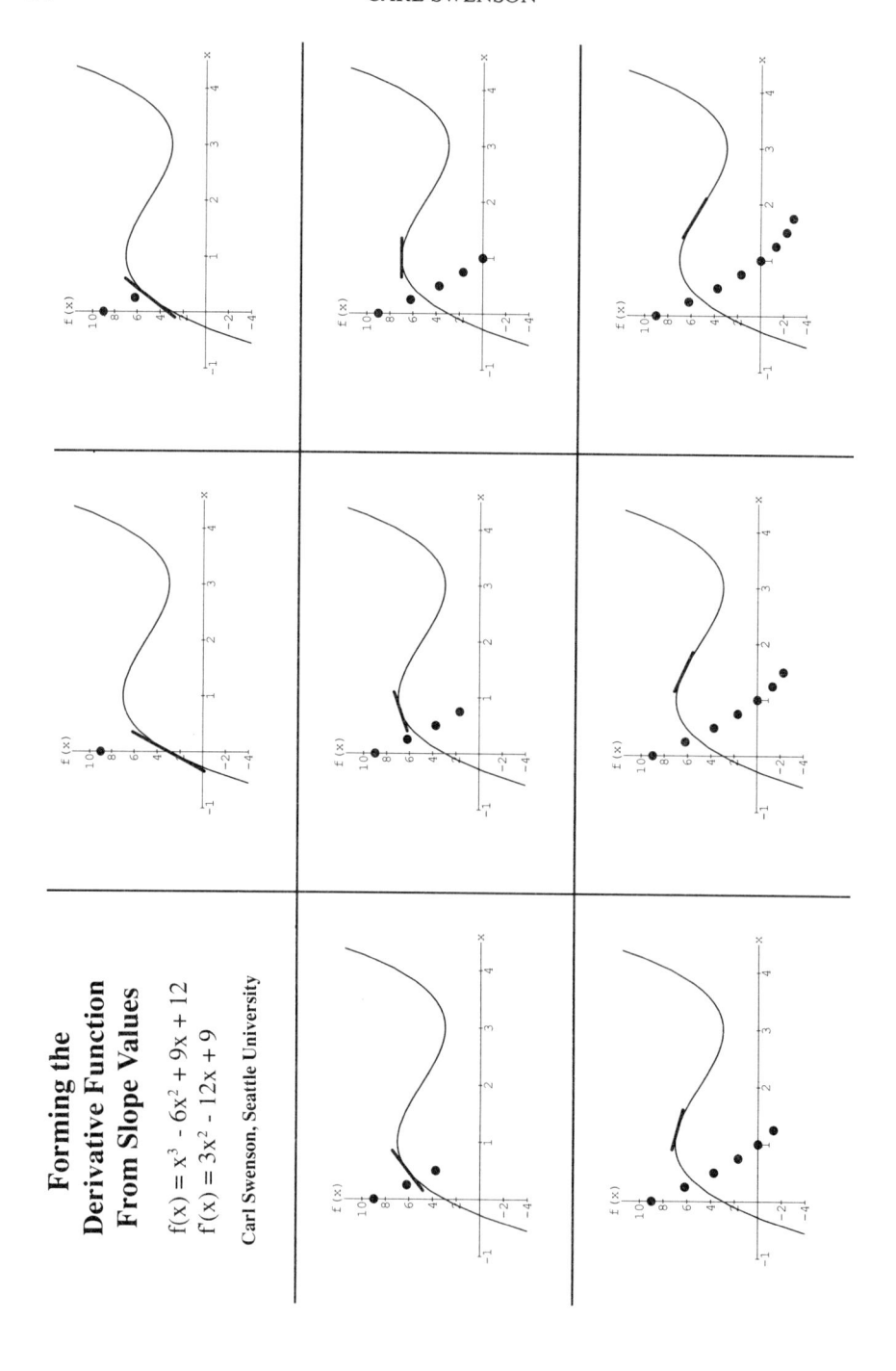

**Forming the Derivative Function From Slope Values**

$f(x) = x^3 - 6x^2 + 9x + 12$
$f'(x) = 3x^2 - 12x + 9$

Carl Swenson, Seattle University

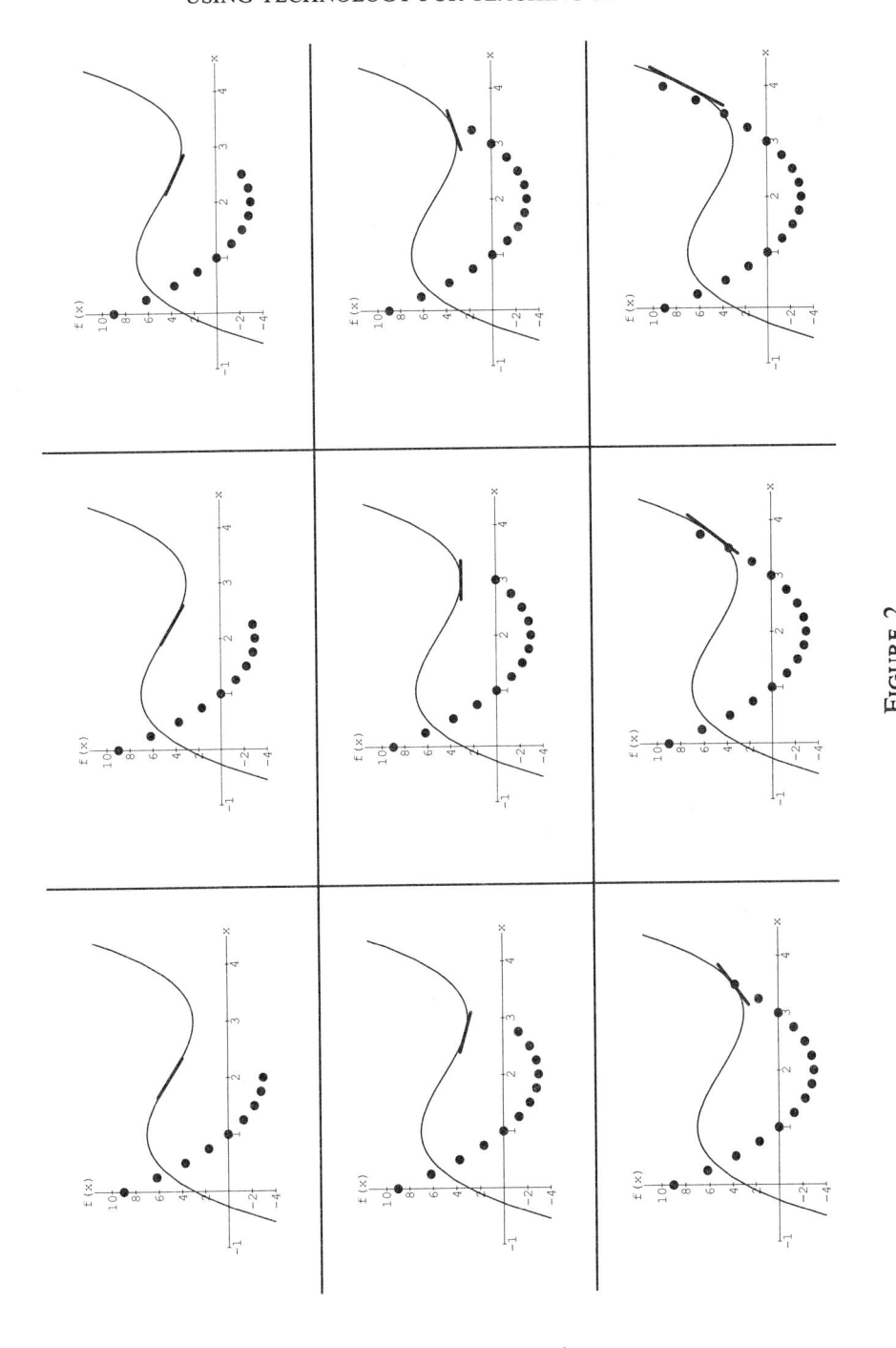

USING TECHNOLOGY FOR TEACHING MATHEMATICS 173

FIGURE 2

verification by statistical analysis—that the attention span of students is growing shorter and that they often expect quick and easy answers. This software does indeed give quick and easy answers, but the road to getting *correct* answers often can be frustratingly long and tedious. The students need to know this.

Criticism has been made that students using computers learn to mindlessly "crank out" answers; but even now students blindly crank out answers by hand. I gave a problem on Valentine's Day to calculate the area of a heart-shaped figure which extended below the $x$-axis. One student using *Mathematica* integrated three parts and then added the integral values together to obtain a negative amount as his answer. Another student used only pencil calculations to obtain a positive value, but one so small as to be clearly incorrect. Neither student saw the unrealistic nature of his answer. Estimating and checking the reasonableness of answers are important.

## Expectations and Results

What can you expect from the introduction of computers into your teaching? At the 1990 *Mathematica* Conference, Dana Scott from Carnegie Mellon said the following about his experimental course in Projective Geometry:

> Mathematically, the most interesting outcome of the experiment was to find out how easy it is to use the ideas of linear space theory, exterior products, and partial differential operators in presenting geometry, both in computations and proofs aided by the power of symbolic computation in *Mathematica*.

By using *Mathematica* I have also found myself working with a broader range of mathematics. It is easy to set up special abstract systems from your own definitions and operations; originally I had not envisioned the kind of algebraic manipulation and proof possible with *Mathematica*.

A short example of a proof, mirrored in Example 5, is to play with an expression, say: $(x+1)(x+2)(x+3)(x+4)+1$ and manipulate to obtain $(x^2+5x+5)^2$, then edit the original to try another case: $(x+2)(x+3)(x+4)(x+5)+1$ and now obtain $(x^2+7x+11)^2$. The glory of *Mathematica* is that I can then enter the constant "$a$" and prove that all such expressions are perfect squares of a quadratic with leading coefficient 1.

$$(x+a)(x+a+1)(x+a+2)(x+a+3)+1$$
$$= (x^2 + (2a+3)x + (a2 + 3a + 10))^2.$$

Of course much deeper work is possible, see, for example, "Exploring Conic Sections with a Computer Algebra System," Lopez and Mathews, *Collegiate Microcomputer*, August 1990.

An article in *The College Mathematics Journal*, January 1990 listed the factorizations (over the real numbers) of $x^n - 1$ for $n$ from 1 to 30. All of

### ■ Finding a pattern and proving a result. (Unknown author)

```
(* Polynomials are output in reverse standard order *)
   Expand[(x+1)(x+2)(x+3)(x+4)+1]

              2      3    4
25 + 50 x + 35 x + 10 x + x

Factor[%]   (* "%" stands for the previous result.*)

        2 2
(5 + 5 x + x )

(* Now I cut & paste and combine to one stage to look again *)
Factor[ Expand[(x+5)(x+2)(x+3)(x+4)+1] ]

         2 2
(11 + 7 x + x )

(* This looks promising so I use a variable "a". *)
Factor[ Expand[(x+a)(x+a+1)(x+a+2)(x+a+3)+1] ]

          2          2 2
(1 + 3 a + a + 3 x + 2 a x + x )
```

This requires human recognition to put it back into a standard form. ( x^2 + (2a + 3)x + (a^2 + 3a + 1) )^2. Thus we have proven that all expressions of the original form are perfect squares of a quadratic with leading coefficent 1.

## EXAMPLE 5

the shown coefficients were either $-1$, $0$, or $1$. It seemed possible that this was generally true, but in a minute, *Mathematica* can find a counterexample at $n = 105$. Imagine the time saved. Using *Mathematica* I tested and gained confidence in certain number theory conjectures concerning Stirling numbers. The calculations used Stirling numbers of fifty or more digits such as the one shown in Example 1. Have I wandered out of mathematics education and into mathematical research? Maybe I have brought mathematical research into mathematics education.

By using technology, realistic applications of mathematics can be used to motivate students (and instructors). The mathematics can be quickly and accurately present in three views: as data, as relation/rule, and as a graphic image. For example, a function definition, a table of values, and a graph of the function can all appear on the screen at the same time. This multiple representation simultaneously accommodates three learning styles: those who prefer detail first, those who prefer theory first, and those who prefer a visual image first. Subsequent understanding of the connection between the multiple representations will deepen student learning.

In summary, the use of technology in mathematics education is surprisingly uncharted territory. Early implementations show very positive results. All teachers of mathematics must incorporate the changes brought by technology. Ignoring technology in the teaching of mathematics is becoming impossible. Avoid the many pitfalls of pioneers, such as getting lost in the glitter of Computer Science, spending endless hours on trivial software or projects, and experiencing hardware/software failure in a presentation. Guard your time and pace yourself for the long run. A clear mathematical or pedagogical purpose should define your use of technology.

DEPARTMENT OF MATHEMATICS, SEATTLE UNIVERSITY, SEATTLE, WASHINGTON 98122